T0291488

LONDON MATHEMATICAL SOCIETY LECTURE NOTE SERIES

Managing Editor: Professor M. Reid, Mathematics Institute, University of Warwick, Coventry CV4 7AL, United Kingdom

The titles below are available from booksellers, or from Cambridge University Press at www.cambridge.org/mathematics

LONDON MATHEMATICAL SOCIETY LECTURE NOTE SERIES: 373

Smoothness, Regularity and Complete Intersection

JAVIER MAJADAS

ANTONIO G. RODICIO

*Universidad de Santiago
de Compostela, Spain*

CAMBRIDGE
UNIVERSITY PRESS

CAMBRIDGE
UNIVERSITY PRESS

University Printing House, Cambridge CB2 8BS, United Kingdom

One Liberty Plaza, 20th Floor, New York, NY 10006, USA

477 Williamstown Road, Port Melbourne, VIC 3207, Australia

314-321, 3rd Floor, Plot 3, Splendor Forum, Jasola District Centre, New Delhi - 110025, India

103 Penang Road, #05-06/07, Visioncrest Commercial, Singapore 238467

Cambridge University Press is part of the University of Cambridge.

It furthers the University's mission by disseminating knowledge in the pursuit of
education, learning and research at the highest international levels of excellence.

www.cambridge.org
Information on this title: www.cambridge.org/9780521125727

© J. Majadas and A. G. Rodicio 2010

First published 2010

A catalogue record for this publication is available from the British Library

ISBN 978-0-521-12572-7 Paperback

Contents

Introduction

This book proves a number of important theorems that are commonly given in advanced books on Commutative Algebra without proof, owing to the difficulty of the existing proofs. In short, we give homological proofs of these results, but instead of the original ones involving simplicial methods, we modify these to use only lower dimensional homology modules, that we can introduce in an ad hoc way, thus avoiding simplicial theory. This allows us to give complete and comparatively short proofs of the important results we state below. We hope these notes can serve as a complement to the existing literature.

These are some of the main results we prove in this book:

Theorem (I) *Let $(A, \mathfrak{m}, K) \to (B, \mathfrak{n}, L)$ be a local homomorphism of noetherian local rings. Then the following conditions are equivalent:*

a) B is a formally smooth A-algebra for the $\mathfrak{n}$-adic topology

b) B is a flat A-module and the K-algebra $B \otimes_A K$ is geometrically regular.

This result is due to Grothendieck [EGA 0_{IV}, (19.7.1)]. His proof is long, though it provides a lot of additional information. He uses this result in proving Cohen's theorems on the structure of complete noetherian local rings. An alternative proof of (I) was given by M. André [An1], based on André–Quillen homology theory; it thus uses simplicial methods, that are not necessarily familiar to all commutative algebraists. A third proof was given by N. Radu [Ra2], making use of Cohen's theorems on complete noetherian local rings.

Theorem (II) *Let A be a complete intersection ring and $\mathfrak{p}$ a prime ideal of A. Then the localization $A_{\mathfrak{p}}$ is a complete intersection.*

1

This result is due to L.L. Avramov [Av1]. Its proof uses differential graded algebras as well as André–Quillen homology modules in dimensions 3 and 4, the vanishing of which characterizes complete intersections.

Our proofs of these two results follow André and Avramov's arguments [An1], [Av1, Av2] respectively, but we make appropriate changes so as to involve André–Quillen homology modules only in dimensions ≤ 2: up to dimension 2 these homology modules are easy to construct following Lichtenbaum and Schlessinger [LS].

Theorem (III) *A regular homomorphism is a direct limit of smooth homomorphisms of finite type (D. Popescu [Po1]–[Po3]).*

We give here Popescu's proof [Po1]–[Po3], [Sw]. An alternative proof is due to Spivakovsky [Sp].

Theorem (IV) *The module of differentials of a regular homomorphism is flat.*

This result follows immediately from (III). However, for many years up to the appearance of Popescu's result, the only known proof was that by André, making essential use of André–Quillen homology modules in *all* dimensions.

Theorem (V) *If $f\colon (A, \mathfrak{m}, K) \to (B, \mathfrak{n}, L)$ is a local formally smooth homomorphism of noetherian local rings and A is quasiexcellent, then f is regular.*

This result is due to André [An2]; we give here a proof more in the style of the methods of this book, mainly following some papers of André, A. Brezuleanu and N. Radu.

We now describe the contents of this book in brief. Chapter 1 introduces homology modules in dimensions 0, 1 and 2. First, in Section 1.1 we give the definition of Lichtenbaum and Schlessinger [LS], which is very concise, at least if we omit the proof that it is well defined. The reader willing to take this on trust and to accept its properties (1.4) can omit Sections (1.2–1.3) on first reading; there, instead of following [LS], we construct the homology modules using differential graded resolutions. This makes the definition somewhat longer, but simplifies the proof of some properties. Moreover, differential graded resolutions are used in an essential way in Chapter 4.

Chapter 2 studies formally smooth homomorphisms, and in particular proves Theorem (I). We follow mainly [An1], making appropriate changes to avoid using homology modules in dimensions > 2. This part was already written (in Spanish) in 1988.

Chapter 3 uses the results of Chapter 2 to deduce Cohen's theorems on complete noetherian local rings. We follow mainly [EGA 0_{IV}] and Bourbaki [Bo, Chapter 9].

In Chapter 4, we prove Theorem (II). After giving Gulliksen's result [GL] on the existence of minimal differential graded resolutions, we follow Avramov [Av1] and [Av2], taking care to avoid homology modules in dimension 3 and 4. As a by-product, we also give a proof of Kunz's result characterizing regular local rings in positive characteristic in terms of the Frobenius homomorphism.

Finally, Chapters 5 and 6 study regular homomorphisms, giving in particular proofs of Theorems (III), (IV) and (V).

The prerequisites for reading this book are a basic course in commutative algebra (Matsumura [Mt, Chapters 1–9] should be more than sufficient) and the first definitions in homological algebra. Though in places we use certain exact sequences deduced from spectral sequences, we give direct proofs of these in the Appendix, thus avoiding the use of spectral sequences.

Finally, we make the obvious remark that this book is not in any way intended as a substitute for André's simplicial homological methods [An1] or the proofs given in [EGA 0_{IV}], since either of these treatments is more complete than ours. Rather, we hope that our book can serve as an introduction and motivation to study these sources. We would also like to mention that we have profited from reading the interesting book by Brezuleanu, Dumitrescu and Radu [BDR] on topics similar to ours, although they do not use homological methods.

We are grateful to T. Sánchet Giralda for interesting suggestions and to the editor for contributing to improve the presentation of these notes.

Conventions. All rings are commutative, except that graded rings are sometimes (strictly) anticommutative; the context should make it clear in each case which is intended.

1

Definition and first properties of (co-)homology modules

In this chapter we define the Lichtenbaum–Schlessinger (co-)homology modules $H_n(A, B, M)$ and $H^n(A, B, M)$, for $n = 0, 1, 2$, associated to a (commutative) algebra $A \to B$ and a B-module M, and we prove their main properties [LS]. In Section 1.1 we give a simple definition of $H_n(A, B, M)$ and $H^n(A, B, M)$, but without justifying that they are in fact well defined. To justify this definition, in Section 1.3 we give another (now complete) definition, and prove that it agrees with that of 1.1. We use differential graded algebras, introduced in Section 1.2. In [LS] they are not used. However we prefer this (equivalent) approach, since we also use differential graded algebras later in studying complete intersections. More precisely, we use Gulliksen's Theorem 4.1.7 on the existence of minimal differential graded algebra resolutions in order to prove Avramov's Lemma 4.2.1. Section 1.4 establishes the main properties of these homology modules.

Note that these (co-)homology modules (defined only for $n = 0, 1, 2$) agree with those defined by André and Quillen using simplicial methods [An1, 15.12, 15.13].

1.1 First definition

Definition 1.1.1 Let A be a ring and B an A-algebra. Let $e_0 \colon R \to B$ be a surjective homomorphism of A-algebras, where R is a polynomial A-algebra. Let $I = \ker e_0$ and

$$0 \to U \to F \xrightarrow{j} I \to 0$$

an exact sequence of R-modules with F free. Let $\phi \colon \bigwedge^2 F \to F$ be the R-module homomorphism defined by $\phi(x \wedge y) = j(x)y - j(y)x$, where $\bigwedge^2 F$

4

is the second exterior power of the R-module F. Let $U_0 = \text{im}(\phi) \subset U$. We have $IU \subset U_0$, and so U/U_0 is a B-module. We have a complex of B-modules

$$U/U_0 \to F/U_0 \otimes_R B = F/IF \to \Omega_{R|A} \otimes_R B$$

(concentrated in degrees 2, 1 and 0), where the first homomorphism is induced by the injection $U \to F$, and the second is the composite $F/IF \to I/I^2 \to \Omega_{R|A} \otimes_R B$, where the first map is induced by j, and the second by the canonical derivation $d \colon R \to \Omega_{R|A}$ (here $\Omega_{R|A}$ is the module of Kähler differentials). We denote any such complex by $\mathbb{L}_{B|A}$, and define for a B-module M

$$H_n(A, B, M) = H_n(\mathbb{L}_{B|A} \otimes_B M) \qquad \text{for } n = 0, 1, 2,$$
$$H^n(A, B, M) = H^n(\text{Hom}_B(\mathbb{L}_{B|A}, M)) \qquad \text{for } n = 0, 1, 2.$$

In Section 1.3 we show that this definition does not depend on the choices of R and F.

1.2 Differential graded algebras

Definition 1.2.1 Let A be a ring. A *differential graded A-algebra* (R, d) (DG A-algebra in what follows) is an (associative) graded A-algebra with unit $R = \bigoplus_{n \geq 0} R_n$, strictly anticommutative, i.e., satisfying

$$xy = (-1)^{pq} yx \text{ for } x \in R_p, y \in R_q \quad \text{and} \quad x^2 = 0 \text{ for } x \in R_{2n+1},$$

and having a differential $d = (d_n \colon R_n \to R_{n-1})$ of degree -1; that is, d is R_0-linear, $d^2 = 0$ and $d(xy) = d(x)y + (-1)^p x d(y)$ for $x \in R_p$, $y \in R$. Clearly, (R, d) is a DG R_0-algebra. We can view any A-algebra B as a DG A-algebra concentrated in degree 0.

A *homomorphism* $f \colon (R, d_R) \to (S, d_S)$ of DG A-algebras is an A-algebra homomorphism that preserves degrees ($f(R_n) \subset S_n$) such that $d_S f = f d_R$.

If (R, d_R), (S, d_S) are DG A-algebras, we define their *tensor product* $R \otimes_A S$ to be the DG A-algebra having

a) underlying A-module the usual tensor product $R \otimes_A S$ of modules, with grading given by

$$R \otimes_A S = \bigoplus_{n \geq 0} \left(\bigoplus_{p+q=n} R_p \otimes_A S_q \right)$$

b) product induced by $(x \otimes y)(x' \otimes y') = (-1)^{pq}(xx' \otimes yy')$ for $y \in S_p$, $x' \in R_q$,

c) differential induced by $d(x \otimes y) = d_R(x) \otimes y + (-1)^q x \otimes d_S(y)$ for $x \in R_q$, $y \in S$.

Let $\{(R_i, d_i)\}_{i \in I}$ be a family of DG A-algebras. For each finite subset $J \subset I$, we extend the above definition to $\bigotimes_{i \in J} {}_A R_i$; for finite subsets $J \subset J'$ of I, we have a canonical homomorphism $\bigotimes_{i \in J} {}_A R_i \to \bigotimes_{i \in J'} {}_A R_i$. We thus have a direct system of homomorphisms of DG A-algebras. We say that the direct limit is the tensor product of the family of DG A-algebras $\{(R_i, d_i)\}_{i \in I}$. It is a DG A-algebra, that we denote by $\bigotimes_{i \in I} {}_A R_i$ (and is not to be confused with the tensor product of the underlying family of A-algebras R_i).

A DG *ideal I* of a DG A-algebra (R, d) is a homogeneous ideal of the graded A-algebra R that is stable under the differential, i.e., $d(I) \subset I$. Then R/I is canonically a DG A-algebra and the canonical map $R \to R/I$ is a homomorphism of DG A-algebras.

An *augmented* DG A-algebra is a DG A-algebra together with a surjective (augmentation) homomorphism of DG A-algebras $p \colon R \to R'$, where R' is a DG A-algebra concentrated in degree 0; its *augmentation ideal* is the DG ideal $\ker p$ of R.

A DG *subalgebra S* of a DG A-algebra (R, d) is a graded A-subalgebra S of R such that $d(S) \subset S$. Let (R, d) be a DG A-algebra. Then $Z(R) := \ker d$ is a graded A-subalgebra of R with grading $Z(R) = \bigoplus_{n \geq 0}(Z(R) \cap R_n)$, and $B(R) := \operatorname{im}(d)$ is a homogeneous ideal of $Z(R)$. Therefore the homology of R

$$H(R) = Z(R)/B(R)$$

is a graded A-algebra.

Example 1.2.2 Let R_0 be an A-algebra and X a variable of degree $n > 0$. Let $R = R_0 \langle X \rangle$ be the following graded A-algebra:

a) If n is odd, $R_0 \langle X \rangle$ is the exterior R_0-algebra on the variable X, i.e., $R_0 \langle X \rangle = R_0 1 \oplus R_0 X$, concentrated in degrees 0 and n.

b) If n is even, $R_0 \langle X \rangle$ is the quotient of the polynomial R_0-algebra on variables $X^{(1)}, X^{(2)}, \ldots$, by the ideal generated by the elements

$$X^{(i)} X^{(j)} - \frac{(i+j)!}{i! j!} X^{(i+j)} \quad \text{for } i, j \geq 1.$$

The grading is defined by $\deg X^{(m)} = nm$ for $m > 0$. We set $X^{(0)} = 1$, $X = X^{(1)}$ and say that $X^{(i)}$ is the ith *divided power* of X. Observe that $i! X^{(i)} = X^i$.

Now let R be a DG A-algebra, x a homogeneous cycle of R of degree $n - 1 \geq 0$, i.e., $x \in Z_{n-1}(R)$. Let X be a variable of degree n, and $R\langle X \rangle = R \otimes_{R_0} R_0 \langle X \rangle$. We define a differential in $R\langle X \rangle$ as the unique differential d for which $R \to R\langle X \rangle$ is a DG A-algebra homomorphism with $d(X) = x$ for n odd, respectively $d(X^{(m)}) = x X^{(m-1)}$ for n even. We denote this DG A-algebra by $R\langle X; dX = x \rangle$.

Note that an augmentation $p \colon R \to R'$ satisfying $p(x) = 0$ extends in a unique way to an augmentation $p \colon R\langle X; dX = x \rangle \to R'$ by setting $p(X) = 0$.

Lemma 1.2.3 *Let R be a DG A-algebra and $c \in H_{n-1}(R)$ for some $n \geq 1$. Let $x \in Z_{n-1}(R)$ be a cycle whose homology class is c. Set $S = R\langle X; dX = x \rangle$ and let $f \colon R \to S$ be the canonical homomorphism. Then:*

a) f induces isomorphisms $H_q(R) = H_q(S)$ for all $q < n - 1$;
b) f induces an isomorphism $H_{n-1}(R)/\langle c \rangle_{R_0} = H_{n-1}(S)$.

Proof a) is clear, since $R_q = S_q$ for $q < n$,
b) $Z_{n-1}(R) = Z_{n-1}(S)$ and $B_{n-1}(R) + x R_0 = B_{n-1}(S)$. □

Definition 1.2.4 If $\{X_i\}_{i \in I}$ is a family of variables of degree > 0, we define $R_0 \langle \{X_i\}_{i \in I} \rangle := \bigotimes_{i \in I} {}_{R_0} R_0 \langle X_i \rangle$ as the tensor product of the DG R_0-algebras $R_0 \langle X_i \rangle$ for $i \in I$ (as in Definition 1.2.1). If R is a DG A-algebra, we say that a DG A-algebra S is *free* over R if the underlying graded A-algebra is of the form $S = R \otimes_{R_0} S_0 \langle \{X_i\}_{i \in I} \rangle$ where S_0 is a polynomial R_0-algebra and $\{X_i\}_{i \in I}$ a family of variables of degree > 0, and the differential of S extends that of R. (Caution: it is not necessarily a free object in the category of DG A-algebras.)

If R is a DG A-algebra and $\{x_i\}_{i \in I}$ a set of homogeneous cycles of R, we define $R\langle \{X_i\}_{i \in I}; dX_i = x_i \rangle$ to be the DG A-algebra

$$R \otimes_{R_0} \left(\bigotimes_{i \in I} {}_{R_0} R_0 \langle X_i; dX_i = x_i \rangle \right),$$

which is free over R.

Lemma 1.2.5 *Let R be a DG A-algebra, $n - 1 \geq 0$, $\{c_i\}_{i \in I}$ a set* *of elements of $H_{n-1}(R)$ and $\{x_i\}_{i \in I}$ a set of homogeneous cycles with classes $\{c_i\}_{i \in I}$. Set $S = R\langle\{X_i\}_{i \in I}; dX_i = x_i\rangle$, and let $f: R \to S$ be the canonical homomorphism. Then:*

  a) f induces isomorphisms $H_q(R) = H_q(S)$ for all $q < n - 1$;
  b) f induces an isomorphism $H_{n-1}(R)/\langle\{c_i\}_{i \in I}\rangle_{R_0} = H_{n-1}(S)$.

Proof Similar to the proof of Lemma 1.2.3, bearing in mind that direct limits are exact. □

Theorem 1.2.6 *Let $p: R \to R'$ be an augmented DG A-algebra. Then there exists an augmented DG A-algebra $p_S: S \to R'$, free over R with $S_0 = R_0$, such that the augmentation p_S extends p and gives an isomorphism in homology*

$$H(S) = H(R') = \begin{cases} R' & \text{if } n = 0, \\ 0 & \text{if } n > 0. \end{cases}$$

If R_0 is a noetherian ring and R_i an R_0-module of finite type for all i, then we can choose S such that S_i is an S_0-module of finite type for all i.

Proof Let $S^0 = R$. Assume that we have constructed an augmented DG A-algebra S^{n-1} that is free over R, such that $S_0^{n-1} = R_0$ and the augmentation $S^{n-1} \to R'$ induces isomorphisms $H_q(S^{n-1}) = H_q(R')$ for $q < n - 1$. Let $\{c_i\}_{i \in I}$ be a set of generators of the R_0-module

$$\ker\left(H_{n-1}(S^{n-1}) \to H_{n-1}(R')\right)$$

(equal to $H_{n-1}(S^{n-1})$ for $n > 1$), and $\{x_i\}_{i \in I}$ a set of homogeneous cycles with classes $\{c_i\}_{i \in I}$. Let $S^n = S^{n-1}\langle\{X_i\}_{i \in I}; dX_i = x_i\rangle$. Then S^n is a DG A-algebra free over R with $S_0^n = R_0$ and such that the augmentation $p_{S^n}: S^n \to R'$ extending $p_{S^{n-1}}$ defined by $p_{S^n}(X_i) = 0$ induces isomorphisms $H_q(S^n) = H_q(R')$ for $q < n$ (Lemma 1.2.5).
  We define $S := \varinjlim S^n$.
  If R_0 is a noetherian ring and R_i an R_0-module of finite type for all i, then by induction we can choose S^n with S_i^n an $S_0^n = R_0$-module of finite type for all i, since if S_i^{n-1} is an S_0^{n-1}-module of finite type for all i, then $H_i(S^{n-1})$ is an S_0^{n-1}-module of finite type for all i. □

Definition 1.2.7 Let $A \to B$ be a ring homomorphism. Let R be a DG A-algebra that is free over A with a surjective homomorphism of DG

A-algebras $R \to B$ inducing an isomorphism in homology. Then we say that R is a *free DG resolution* of the A-algebra B.

Corollary 1.2.8 *Let $A \to B$ be a ring homomorphism. Then a free DG resolution R of the A-algebra B exists. If A is noetherian and B an A-algebra of finite type, then we can choose R such that R_0 is a polynomial A-algebra of finite type and R_i an R_0-module of finite type for all i.*

Proof Let R_0 be a polynomial A-algebra such that there exists a surjective homomorphism of A-algebras $R_0 \to B$. (If A is noetherian and B an A-algebra of finite type, then we can choose R_0 a polynomial A-algebra of finite type.) Now apply Theorem 1.2.6 to $R_0 \to B$. $\square$

Definition 1.2.9 Let R be a DG A-algebra that is free over R_0, i.e., $R = R_0 \langle \{X_i\}_{i \in I} \rangle$. For $n \geq 0$, we define the *n-skeleton* of R to be the DG R_0-subalgebra generated by the variables X_i of degree $\leq n$ and their divided powers (for variables of even degree > 0). We denote it by $R(n)$. Thus $R(0) = R_0$, and if $A \to B$ is a surjective ring homomorphism with kernel I and R a free DG resolution of the A-algebra B with $R_0 = A$, then $R(1)$ is the Koszul complex associated to a set of generators of I.

Lemma 1.2.10 *Let A be a ring and B an A-algebra. Let*

$$
\begin{array}{ccc}
A & \to & S \\
\downarrow & \nearrow & \downarrow \\
R & \to & B
\end{array}
$$

be a commutative diagram of DG A-algebra homomorphisms, where S is a free DG resolution of the S_0-algebra B and R is a DG A-algebra that is free over A. Then there exists a DG A-algebra homomorphism $R \to S$ that makes the whole diagram commute.

Proof Let $R(n)$ be the n-skeleton of R. Assume by induction that we have defined a homomorphism of DG A-algebras $R(n-1) \to S$ so that the associated diagram commutes. We extend it to a DG A-algebra homomorphism $R(n) \to S$ keeping the commutativity of the diagram.

   a) If $n = 0$, $R(0) = R_0$ and $R_0 \to S_0$ exists because R_0 is a polynomial A-algebra.

b) If n is odd, let $R(n) = R(n-1)\langle\{T_i\}_{i\in I}\rangle$. We have a commutative diagram

$$
\begin{array}{ccccc}
R(n-1)_n \oplus \bigoplus_{i\in I} R_0 T_i & R(n-1)_{n-1} & & R(n-1)_{n-2} \\
\| & \| & & \| \\
R(n)_n & \longrightarrow & R(n)_{n-1} & \longrightarrow & R(n)_{n-2} \\
& & \downarrow & & \downarrow \\
S_n & \longrightarrow & S_{n-1} & \longrightarrow & S_{n-2}
\end{array}
$$

and therefore a homomorphism $R(n)_n \to \ker(S_{n-1} \to S_{n-2}) = \operatorname{im}(S_n \to S_{n-1})$, and so there exist an R_0-module homomorphism $R(n)_n \to S_n$ extending $R(n-1)_n \to S_n$. By multiplicativity using the map $R(n)_n \to S_n$, we extend $R(n-1) \to S$ to a homomorphism of DG A-algebras $R(n) \to S$.

c) For even $n \geq 2$, suppose that $R(n) = R(n-1)\langle\{X_i\}_{i\in I}\rangle$. As above, we define $R(n)_n \to S_n$ and then extend it to $R(n) \to S$ by multiplicativity using divided power rules based on the binomial and multinomial theorems.

In more detail, suppose the map $R(n)_n \to S_n$ is defined by

$$
X_i \mapsto \sum_{t=1}^{v} a_t Y_1^{(r_{t,1})} \cdots Y_m^{(r_{t,m})} \in S_n,
$$

where the a_t are coefficients in S_0, the Y_i are variables with $\deg Y_i > 0$, and the divided powers $Y_j^{(r_{t,j})}$ have integer exponents $r_{t,j} \geq 0$. (Of course, for $\deg Y_j$ odd and $r > 1$, we understand $Y_j^{(r)} = 0$.) Then for $l > 0$, the image of $X_i^{(l)}$ is determined by the familiar divided power rules†

(a) $(Y_1 + \cdots + Y_v)^{(l)} = \displaystyle\sum_{\substack{\alpha_1+\cdots+\alpha_v=l \\ \alpha_1,\ldots,\alpha_v \geq 0}} Y_1^{(\alpha_1)} \cdots Y_v^{(\alpha_v)}$; and

(b) $(Y_1 Y_2)^{(l)} = Y_1^l Y_2^{(l)}$ (if $\deg Y_1$ and $\deg Y_2 \geq 2$ are even).

Thus $R(n) \to S_n$ is given by

$$
X_i^{(l)} \mapsto \sum_{\substack{\alpha_1+\cdots+\alpha_v=l \\ \alpha_1,\ldots,\alpha_v \geq 0}} \left(\prod_{t=1}^{v} a_t^{\alpha_t} \frac{(Y_1^{(r_{t,1})})^{\alpha_t} \cdots (Y_m^{(r_{t,m})})^{\alpha_t}}{\alpha_t!} \right),
$$

† Both are justified by observing that the two sides agree on multiplying by $l!$.

where the monomial

$$\frac{(Y_1^{(r_{t,1})})^{\alpha_t} \cdots (Y_m^{(r_{t,m})})^{\alpha_t}}{\alpha_t!}$$

equals

- 1 if $\alpha_t = 0$;
- 0 if $\alpha_t \geq 2$ and $r_{t,j} = 0$ for every j with deg Y_j even positive;
- $\frac{(r_{t,j}\alpha_t)!}{\alpha_t!(r_{t,j}!)^{\alpha_t}} \times (Y_1^{(r_{t,1})})^{\alpha_t} \cdots Y_j^{(r_{t,j}\alpha_t)} \cdots (Y_m^{(r_{t,m})})^{\alpha_t}$ if $\alpha_t = 1$, or if for some j deg Y_j is even and positive and $r_{t,j}\alpha_t \geq 1$;

note that the coefficient $\frac{(r_{t,j}\alpha_t)!}{\alpha_t!(r_{t,j}!)^{\alpha_t}}$ is an integer.

Using the formula $Y_i^{(p)}Y_i^{(q)} = \frac{(p+q)!}{p!q!}Y_i^{(p+q)}$, we see that $(Y_i^{(r_{t,i})})^{\alpha_t} = \frac{(r_{t,i}\alpha_t)!}{(r_{t,i}!)^{\alpha_t}}Y_i^{(r_{t,i}\alpha_t)}$, and so this definition does not depend on the chosen j.

A straightforward computation (easier if we multiply "formally" by $p!q!$), shows that under this map, $X_i^{(p)}X_i^{(q)}$ and $\frac{(p+q)!}{p!q!}X_i^{(p+q)}$ have the same image. $\qquad\square$

Remarks

i) The assumption that S is free over S_0 is only used to avoid defining divided powers structure.

ii) For the definition of $H_n(A, B, M)$, for $n = 0, 1, 2$, we use free DG resolutions only up to degree 3, and so we could have used symmetric powers resolutions instead of divided powers resolutions (since they agree in degrees ≤ 3). However, in Chapter 4 we use minimal resolutions and there we need divided powers.

1.3 Second definition

Definition 1.3.1 Let $A \to B$ be a ring homomorphism. Let $e\colon R \to B$ be a free DG resolution of the A-algebra B. Let $J = \ker(R \otimes_A B \to B,$ $x \otimes b \mapsto e(x)b)$. Let $J^{(2)}$ be the graded $R_0 \otimes_A B$-submodule of $R \otimes_A B$ generated by the products of the elements of J and the divided powers $X^{(m)}$, $m > 1$ of variables of J of even degree ≥ 2. Note that $J^{(2)}$ is a subcomplex of $R_0 \otimes_A B$-modules of J. We define the complex

$$\Omega_{R|A} \otimes_R B := J/J^{(2)},$$

which is in fact a complex of B-modules.

In degree 0 it is isomorphic to $\Omega_{R_0|A} \otimes_{R_0} B$, where $\Omega_{R_0|A}$ is the usual

R_0-module of differentials of the A-algebra R_0. For, we have an exact sequence of R_0-modules defined by the multiplication of R_0 (considering $R_0 \otimes_A R_0$ as an R_0-module multiplying in the right factor)

$$0 \to I \to R_0 \otimes_A R_0 \to R_0 \to 0,$$

which splits, and so applying $- \otimes_{R_0} B$ we obtain an exact sequence

$$0 \to I \otimes_{R_0} B \to R_0 \otimes_A B \to B \to 0,$$

showing that $I \otimes_{R_0} B = J_0$. On the other hand, the exact sequence of R_0-modules

$$0 \to I^2 \to I \to \Omega_{R_0|A} \to 0$$

gives an exact sequence

$$I^2 \otimes_{R_0} B = (I \otimes_{R_0} B)^2 = J_0^2 \to I \otimes_{R_0} B = J_0 \to \Omega_{R_0|A} \otimes_{R_0} B \to 0,$$

and therefore $J_0/J_0^{(2)} = J_0/J_0^2 = \Omega_{R_0|A} \otimes_{R_0} B$.

In degree 1, $(J/J^{(2)})_1 = J_1/J_0 J_1 = (R_1 \otimes_A B)/J_0(R_1 \otimes_A B) = (R_1 \otimes_A B) \otimes_{R_0 \otimes_A B} B = R_1 \otimes_{R_0} B$ is the free B-module obtained by base extension of the free R_0-module R_1.

Similarly, in degree 2, $(J/J^{(2)})_2 = J_2/(J_0 J_2 + J_1^2) = (R_2/R_1^2) \otimes_{R_0} B$.

In general, for $n > 0$, $(\Omega_{R|A} \otimes_R B)_n = (R(n)/R(n-1))_n \otimes_{R_0} B$.

Definition 1.3.2 We say that an A-algebra P has property (L) if for any A-algebra Q, any Q-module M, any Q-module homomorphism $u \colon M \to Q$ such that $u(x)y = u(y)x$ for all $x, y \in M$, and for any pair of A-algebra homomorphisms $f, g \colon P \to Q$ such that $\mathrm{im}(f - g) \subset \mathrm{im}(u)$, there exists a biderivation $\lambda \colon P \to M$ such that $u\lambda = f - g$

$$\begin{array}{ccc} & & P \\ & \swarrow^{\lambda} & {\scriptstyle f} \big\Vert {\scriptstyle g} \\ M & \xrightarrow{\ u\ } & Q. \end{array}$$

Here we say that λ is a biderivation to mean that λ is A-linear and $\lambda(xy) = f(x)\lambda(y) + g(y)\lambda(x)$.

Lemma 1.3.3 *Let A be a ring, P an A-algebra.*

  a) *If P is a polynomial A-algebra, then P has property (L).*
  b) *If P has property (L) and S is a multiplicative subset of P, then $S^{-1}P$ has property (L).*

Proof a) Let Q, M, u, f, g be as in (1.3.2). Let $P = A[\{X_i\}_{i \in I}]$. Since $\mathrm{im}(f - g) \subset \mathrm{im}(u)$, there exist elements Y_i in M such that $u(Y_i) = f(X_i) - g(X_i)$. Define λ on monomials $X_{i_1} \cdots X_{i_n}$ by

$$\lambda(X_{i_1} \cdots X_{i_n}) = \sum_{j=1}^{n} f(X_{i_1} \cdots X_{i_{j-1}}) Y_{i_j} g(X_{i_{j+1}} \cdots X_{i_n}).$$

To see that λ is well defined, we show that the right-hand side of this formula remains invariant under transpositions (i_p, i_{p+1}) of the indices:

$$\begin{aligned}
&\lambda(X_{i_1} \cdots X_{i_{p+1}} X_{i_p} \cdots X_{i_n}) - \lambda(X_{i_1} \cdots X_{i_p} X_{i_{p+1}} \cdots X_{i_n}) \\
&= f(X_{i_1} \cdots X_{i_{p-1}}) Y_{i_{p+1}} g(X_{i_p} X_{i_{p+2}} \cdots X_{i_n}) \\
&\quad + f(X_{i_1} \cdots X_{i_{p-1}} X_{i_{p+1}}) Y_{i_p} g(X_{i_{p+2}} \cdots X_{i_n}) \\
&\quad - f(X_{i_1} \cdots X_{i_{p-1}}) Y_{i_p} g(X_{i_{p+1}} \cdots X_{i_n}) \\
&\quad - f(X_{i_1} \cdots X_{i_p}) Y_{i_{p+1}} g(X_{i_{p+2}} \cdots X_{i_n}) \\
&= f(X_{i_1} \cdots X_{i_{p-1}}) g(X_{i_{p+2}} \cdots X_{i_n})(u(x)y - u(y)x) = 0,
\end{aligned}$$

where $x = Y_{i_{p+1}}$ and $y = Y_{i_p}$. Extend λ to P by linearity. It follows easily that λ is a biderivation. Since $f - g$ is also a biderivation and $f(X_i) - g(X_i) = u(Y_i) = u(\lambda(X_i))$, we have $u\lambda = f - g$.

b) Let Q be an A-algebra and M a Q-module and $u \colon M \to Q$ a Q-module homomorphism such that $u(x)y = u(y)x$ for all $x, y \in M$; suppose that $f', g' \colon S^{-1}P \to Q$ are two A-algebra homomorphisms such that $\mathrm{im}(f' - g') \subset \mathrm{im}(u)$. Let $f, g \colon P \to Q$ be the respective composites of f', g' with the canonical map $P \to S^{-1}P$. Since P has property (L), there exists a biderivation $\lambda \colon P \to M$ such that $u\lambda = f - g$.

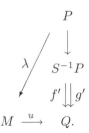

Let $p/s \in S^{-1}P$ with $p \in P$ and $s \in S$. Solving the equation $\lambda(p) = \lambda(s \cdot p/s) = f'(s)\lambda'(p/s) + g'(p/s)\lambda(s)$ for λ' gives

$$\lambda'(p/s) = (g'(s)\lambda(p) - g'(p)\lambda(s))/(f'(s)g'(s)).$$

Using the relation $u(x)y = u(y)x$ for $x, y \in M$, one can show by a tedious but straightforward calculation that this formula actually defines a biderivation $\lambda': S^{-1}P \to M$ extending λ. Since $f' - g'$ is a biderivation extending $f - g$, it is clear that $u\lambda' = f' - g'$. $\square$

Lemma 1.3.4 *Let $A \to B$ be a ring homomorphism and $p: R \to B$ and $q: S \to B$ two free DG resolutions of the A-algebra B. Let $f, g: R \to S$ be two homomorphisms of augmented DG A-algebras, i.e., $p = qf$, $p = qg$. Then there exist B-module homomorphisms*

$$\overline{\alpha}_i: (\Omega_{R|A} \otimes_R B)_i \to (\Omega_{S|A} \otimes_S B)_{i+1} \qquad \text{for } i = 0, 1, 2$$

such that $\overline{d_1^S \alpha_0} = \overline{f}_0 - \overline{g}_0$, and $\overline{d_{i+1}^S \alpha_i} + \overline{\alpha_{i-1} d_i^R} = \overline{f}_i - \overline{g}_i$, for $i = 1, 2$, where d^R, d^S are the differentials of R and S respectively, and $^-$ denotes the induced map in the following diagram

$$(\Omega_{R|A} \otimes_R B)_3 \xrightarrow{\overline{d_3^R}} (\Omega_{R|A} \otimes_R B)_2 \xrightarrow{\overline{d_2^R}} (\Omega_{R|A} \otimes_R B)_1 \xrightarrow{\overline{d_1^R}} (\Omega_{R|A} \otimes_R B)_0$$

$$\overline{f}_3 \Big\| \overline{g}_3 \quad \overset{\overline{\alpha}_2}{\nearrow} \quad \overline{f}_2 \Big\| \overline{g}_2 \quad \overset{\overline{\alpha}_1}{\nearrow} \quad \overline{f}_1 \Big\| \overline{g}_1 \quad \overset{\overline{\alpha}_0}{\nearrow} \quad \overline{f}_0 \Big\| \overline{g}_0$$

$$(\Omega_{S|A} \otimes_S B)_3 \xrightarrow{\overline{d_3^S}} (\Omega_{S|A} \otimes_S B)_2 \xrightarrow{\overline{d_2^S}} (\Omega_{S|A} \otimes_S B)_1 \xrightarrow{\overline{d_1^S}} (\Omega_{S|A} \otimes_S B)_0.$$

Proof Let $I = \ker(R \otimes_A B \to B)$, $J = \ker(S \otimes_A B \to B)$, so that $\Omega_{R|A} \otimes_R B = I/I^{(2)}$, $\Omega_{S|A} \otimes_S B = J/J^{(2)}$. We have $f_i(I_i) \subset J_i$, $g_i(I_i) \subset J_i$, for all i. We begin by defining maps $\widetilde{\alpha}_i$

$$R_3 \xrightarrow{d_3^R} R_2 \xrightarrow{d_2^R} R_1 \xrightarrow{d_1^R} R_0$$

$$\widetilde{f}_3 \Big\| \widetilde{g}_3 \quad \overset{\widetilde{\alpha}_2}{\nearrow} \quad \widetilde{f}_2 \Big\| \widetilde{g}_2 \quad \overset{\widetilde{\alpha}_1}{\nearrow} \quad \widetilde{f}_1 \Big\| \widetilde{g}_1 \quad \overset{\widetilde{\alpha}_0}{\nearrow} \quad \widetilde{f}_0 \Big\| \widetilde{g}_0$$

$$S_3 \xrightarrow{d_3^S} S_2/S_1^2 \xrightarrow{d_2^S} S_1/d_2^S(S_1^2) \xrightarrow{d_1^S} S_0$$

and the $\overline{\alpha}_i$ are the maps induced by the $\widetilde{\alpha}_i$. Note that the lower row is still exact.

Definition of $\widetilde{\alpha}_0$: by Lemma 1.3.3, there exists a biderivation $\widetilde{\alpha}_0: R_0 \to S_1/d_2^S(S_1^2)$ such that $\widetilde{d_1^S \alpha_0} = \widetilde{f}_0 - \widetilde{g}_0$. The A-linear map $\widetilde{\alpha}_0$ induces a map $R_0 \otimes_A B \to S_1/d_2^S(S_1^2) \otimes_A B \to S_1/d_2^S(S_1^2) \otimes_{S_0} B = S_1 \otimes_{S_0} B = (\Omega_{S|A} \otimes_S B)_1$, and this composite factors through a map $\overline{\alpha}_0: (\Omega_{R|A} \otimes_R B)_0 \to (\Omega_{S|A} \otimes_S B)_1$, since $f_0(I_0) \subset J_0$, $g_0(I_0) \subset J_0$, $\widetilde{\alpha}_0$ is a biderivation and then $\widetilde{\alpha}_0 \otimes_A B$ takes I_0^2 into the image of $J_0 J_1$.

Definition of $\widetilde{\alpha}_1$: let $R_1 = \bigoplus_i R_0 T_i$. For each i, let $y_i \in S_2/S_1^2$ be such that $\widetilde{d}_2^S(y_i) = \widetilde{f}_1(T_i) - \widetilde{g}_1(T_i) - \widetilde{\alpha}_0 d_1^R(T_i)$; such a y_i exists, since

$$\widetilde{d}_1^S(\widetilde{f}_1(T_i) - \widetilde{g}_1(T_i) - \widetilde{\alpha}_0 d_1^R(T_i)) =$$
$$\widetilde{f}_0 d_1^R(T_i) - \widetilde{g}_0 d_1^R(T_i) - (\widetilde{f}_0 - \widetilde{g}_0)(d_1^R(T_i)) = 0.$$

Define $\widetilde{\alpha}_1(T_i) = y_i$ and extend it to R_1 by R_0-linearity (regarding S_2 as an R_0-module via f_0). It is clear that $\widetilde{\alpha}_1$ induces $\overline{\alpha_1} \colon (\Omega_{R|A} \otimes_R B)_1 \to (\Omega_{S|A} \otimes_S B)_2$, since $(\widetilde{f}_0 \otimes_A B)(I_0) \subset J_0$. By A-linearity, to see that

$$\widetilde{d}_2^S \widetilde{\alpha}_1 + \widetilde{\alpha}_0 d_1^R = \widetilde{f}_1 - \widetilde{g}_1$$

it is enough to see that

$$\widetilde{d}_2^S \widetilde{\alpha}_1(aT_i) + \widetilde{\alpha}_0 d_1^R(aT_i) = \widetilde{f}_1(aT_i) - \widetilde{g}_1(aT_i)$$

with $a \in R_0$. But

$$\widetilde{d}_2^S \widetilde{\alpha}_1(aT_i) + \widetilde{\alpha}_0 d_1^R(aT_i) - \left(\widetilde{f}_1(aT_i) - \widetilde{g}_1(aT_i)\right)$$
$$= \widetilde{f}_0(a)\widetilde{d}_2^S \widetilde{\alpha}_1(T_i) + \left(\widetilde{f}_0(a)\widetilde{\alpha}_0 d_1^R(T_i) + \widetilde{g}_0(d_1^R(T_i))\widetilde{\alpha}_0(a)\right)$$
$$\quad - \left(\widetilde{f}_0(a)\widetilde{f}_1(T_i) - \widetilde{g}_0(a)\widetilde{g}_1(T_i)\right)$$
$$= \widetilde{f}_0(a)\left(\widetilde{d}_2^S \widetilde{\alpha}_1(T_i) + \widetilde{\alpha}_0 d_1^R(T_i) - \left(\widetilde{f}_1(T_i) - \widetilde{g}_1(T_i)\right)\right)$$
$$\quad + \widetilde{g}_0(d_1^R(T_i))\widetilde{\alpha}_0(a) - \left(\widetilde{f}_0(a) - \widetilde{g}_0(a)\right)(\widetilde{g}_1(T_i))$$
$$= \widetilde{f}_0(a)\left(\widetilde{d}_2^S \widetilde{\alpha}_1(T_i) + \widetilde{\alpha}_0 d_1^R(T_i) - \left(\widetilde{f}_1(T_i) - \widetilde{g}_1(T_i)\right)\right)$$
$$\quad + \widetilde{d}_1^S(\widetilde{g}_1(T_i))\widetilde{\alpha}_0(a)) - \widetilde{d}_1^S(\widetilde{\alpha}_0(a))\widetilde{g}_1(T_i)$$
$$= \widetilde{f}_0(a)\left(\widetilde{d}_2^S \widetilde{\alpha}_1(T_i) + \widetilde{\alpha}_0 d_1^R(T_i) - \left(\widetilde{f}_1(T_i) - \widetilde{g}_1(T_i)\right)\right)$$
$$\quad + \widetilde{d}_2^S(\widetilde{g}_1(T_i)\widetilde{\alpha}_0(a)),$$

where the products $\widetilde{g}_1(T_i)\widetilde{\alpha}_0(a)$ have the obvious meaning. Now the two summands are zero, the first by definition of $\widetilde{\alpha}_1$ and the second because it is in $d_2^S(S_1^2)$.

Definition of $\widetilde{\alpha}_2$: let $R_2 = (\bigoplus_k R_0 U_k) \oplus (\bigoplus_{i \neq j} R_0 T_i T_j)$. By the equation $\widetilde{d}_2^S \widetilde{\alpha}_1 + \widetilde{\alpha}_0 d_1^R = \widetilde{f}_1 - \widetilde{g}_1$, there exist $z_k \in S_3$ such that $\widetilde{d}_3^S(z_k) = \widetilde{f}_2(U_k) - \widetilde{g}_2(U_k) - \widetilde{\alpha}_1 d_2^R(U_k)$. Define $\widetilde{\alpha}_2(U_k) = z_k$. Define $\widetilde{\alpha}_2(T_i T_j) \in S_3$ such that its image in $(\Omega_{S|A} \otimes_S B)_3$ is 0, e.g., define $\widetilde{\alpha}_2(T_i T_j) = 0$ or $\widetilde{\alpha}_2(T_i T_j) = \widetilde{f}_1(T_i)\widetilde{\alpha}_1(T_j) - \widetilde{f}_1(T_j)\widetilde{\alpha}_1(T_i)$ (this second definition has the following advantage, here unnecessary since we do not need $\widetilde{\alpha}_2$, but only $\overline{\alpha_2} \colon (\widetilde{d}_3^S \widetilde{\alpha}_2 + \widetilde{\alpha}_1 d_2^R)(T_i T_j) = 0 = \widetilde{f}_2(T_i T_j) - \widetilde{g}_2(T_i T_j))$. Extend $\widetilde{\alpha}_2$ to R_2 by R_0-linearity (again regarding S_3 as an R_0-module via f_0). As for

$\widetilde{\alpha}_1$, it is clear that $\widetilde{\alpha}_2$ induces $\overline{\alpha_2}\colon (\Omega_{R|A} \otimes_R B)_2 \to (\Omega_{S|A} \otimes_S B)_3$. The equation

$$\overline{d_3^S}\,\overline{\alpha_2} + \overline{\alpha_1}\,\overline{d_2^R} = \overline{f_2} - \overline{g_2}$$

follows from $\widetilde{d_3^S}\,\widetilde{\alpha}_2(U_k) = \widetilde{d_3^S}(z_k) = \widetilde{f}_2(U_k) - \widetilde{g}_2(U_k) - \widetilde{\alpha}_1 d_2^R(U_k)$, and from the R_0-linearity of $\widetilde{d_3^S}$, d_2^R, $\widetilde{\alpha}_2$, $\widetilde{\alpha}_1$ and of $\overline{f_2} - \overline{g_2}$. □

Remark 1.3.5 Again, the assumption that S is free is not necessary.

Definition 1.3.6 Let $A \to B$ be a ring homomorphism and M a B-module. Let R be a free DG resolution of the A-algebra B. For $n = 0, 1, 2$, define the (co-)homology B-modules by

$$H_n(A, B, M) = H_n(\Omega_{R|A} \otimes_R M) := H_n((\Omega_{R|A} \otimes_R B) \otimes_B M)$$

$$H^n(A, B, M) = H^n(\mathrm{Hom}_B(\Omega_{R|A} \otimes_R B, M)).$$

In view of Lemma 1.2.10 and Lemma 1.3.4, the definition does not depend on the choice of R, and is natural in A, B and M.

Proposition 1.3.7 *The (co-)homology modules of Definition 1.3.6 agree with those of Definition 1.1.1, which are therefore well defined.*

Proof Let A be a ring and B an A-algebra. Let $e_0\colon R \to B$ be a surjective homomorphism of A-algebras, where R is a polynomial A-algebra. Let $I = \ker(e_0)$ and

$$0 \to U \to F \xrightarrow{j} I \to 0$$

an exact sequence of R-modules with F free. Let $\phi\colon \bigwedge^2 F \to F$ be as in Definition 1.1.1, and $U_0 = \mathrm{im}(\phi)$.

 Let S be a free DG resolution of the A-algebra B with $S_0 = R$ (see the proof of corollary (1.2.8)) and with $S_1 = F$ as $S_0 = R$-module (see the proof of Theorem 1.2.6).

 We prove that there is an isomorphism between the complex

$$U/U_0 \to F/IF \to \Omega_{R|A} \otimes_R B$$

of Definition 1.1.1, and

$$(\Omega_{S|A} \otimes_S B)_2/d_3(\Omega_{S|A} \otimes_S B)_3 \to (\Omega_{S|A} \otimes_S B)_1 \to (\Omega_{S|A} \otimes_S B)_0.$$

Since $H_n(\Omega_{S|A} \otimes_S B)$ for $n = 0, 1, 2$, do not depend on the choice of S, this proves the proposition (note also that

$$(\Omega_{S|A} \otimes_S B)_2/d_3(\Omega_{S|A} \otimes_S B)_3 \otimes_B M = (\Omega_{S|A} \otimes_S M)_2/d_3(\Omega_{S|A} \otimes_S M)_3,$$

$\mathrm{Hom}_B((\Omega_{S|A} \otimes_S B)_2/d_3(\Omega_{S|A} \otimes_S B)_3, M) =$
$\ker(\mathrm{Hom}_B((\Omega_{S|A} \otimes_S B)_2, M) \to \mathrm{Hom}_B((\Omega_{S|A} \otimes_S B)_3, M))$, for any B-module M).

It is clear that $(\Omega_{S|A} \otimes_S B)_0 = \Omega_{R|A} \otimes_R B$, $(\Omega_{S|A} \otimes_S B)_1 = S_1 \otimes_{S_0} B = F \otimes_R B = F/IF$. To see that $U/U_0 = (\Omega_{S|A} \otimes_S B)_2/d_3(\Omega_{S|A} \otimes_S B)_3$, first note that U/U_0 is the first Koszul homology module of the ideal I of R, that is, $U/U_0 = H_1(S(1))$, where $S(n)$ is the n-skeleton of S.

Associated to the exact sequence

$$0 \to S(1) \to S(2) \to S(2)/S(1) \to 0$$

we have an exact sequence

$$H_2(S(2)) \xrightarrow{\beta_2} H_2(S(2)/S(1)) \xrightarrow{\delta_2} H_1(S(1)) \xrightarrow{\alpha_1} H_1(S(2)) = 0$$

and so $U/U_0 = H_1(S(1)) = H_2(S(2)/S(1))/\mathrm{im}(\beta_2)$.

We have isomorphisms $H_n(S(n)/S(n-1)) = (S(n)/S(n-1))_n \otimes_{S_0} B = (\Omega_{S|A} \otimes_S B)_n$ for $n > 1$, and with this identification, the diagram

$$
\begin{array}{ccc}
H_3(S(3)/S(2)) & \xrightarrow{\delta_3} & H_2(S(2)) & \xrightarrow{\beta_2} & H_2(S(2)/S(1)) \\
\| & & & & \| \\
(\Omega_{S|A} \otimes_S B)_3 & \xrightarrow{d_3} & & & (\Omega_{S|A} \otimes_S B)_2
\end{array}
$$

is commutative. On the other hand, $H_2(S(3)) = 0$, so that $\alpha_2 = 0$ and then δ_3 is surjective. Thus $(\Omega_{S|A} \otimes_S B)_2/d_3(\Omega_{S|A} \otimes_S B)_3 = H_2(S(2)/S(1))/\mathrm{im}(\beta_2) = U/U_0$. So we have vertical isomorphisms in the diagram

$$
\begin{array}{ccccc}
U/U_0 & \longrightarrow & F/IF & \longrightarrow & \Omega_{R|A} \otimes_R B \\
\| & & \| & & \| \\
(\Omega_{S|A} \otimes_S B)_2/d_3(\Omega_{S|A} \otimes_S B)_3 & \to & (\Omega_{S|A} \otimes_S B)_1 & \to & (\Omega_{S|A} \otimes_S B)_0
\end{array}
$$

and it is an easy computation to see that the diagram commutes. $\square$

1.4 Main properties

Proposition 1.4.1 *Let A be a ring, B an A-algebra, M a B-module. We have*

a) $H_0(B, B, M) = 0 = H^0(B, B, M)$.
b) $H_0(A, B, M) = \Omega_{B|A} \otimes_B M$,
   $H^0(A, B, M) = \mathrm{Hom}_B(\Omega_{B|A}, M) = \mathrm{Der}_A(B, M)$ *(A-derivations from B to M)*.
c) *If* $B = A/I$ *where* I *is an ideal of* A, $H_1(A, B, M) = I/I^2 \otimes_B M$, $H^1(A, B, M) = \mathrm{Hom}_B(I/I^2, M)$.
d) *If* B *is a polynomial* A*-algebra,* $H_n(A, B, M) = 0 = H^n(A, B, M)$ *for* $n = 1, 2$.

Proof All the properties follow easily from Definition 1.1.1.   □

Proposition 1.4.2 *Let* A *be a ring,* B *an* A*-algebra and*

$$0 \to M' \to M \to M'' \to 0$$

an exact sequence of B*-modules. Then there exist exact sequences*

$$\begin{aligned}
H_2(A, B, M') &\to H_2(A, B, M) \to H_2(A, B, M'') \to \\
H_1(A, B, M') &\to H_1(A, B, M) \to H_1(A, B, M'') \to \\
H_0(A, B, M') &\to H_0(A, B, M) \to H_0(A, B, M'') \to 0
\end{aligned}$$

and

$$\begin{aligned}
0 \to H^0(A, B, M') &\to H^0(A, B, M) \to H^0(A, B, M'') \to \\
H^1(A, B, M') &\to H^1(A, B, M) \to H^1(A, B, M'') \to \\
H^2(A, B, M') &\to H^2(A, B, M) \to H^2(A, B, M''). \quad □
\end{aligned}$$

Proposition 1.4.3 (Base change) *Let* $A \to B$ *and* $A \to C$ *be ring homomorphisms such that* $\mathrm{Tor}_i^A(B, C) = 0$ *for* $i = 1, 2$. *Let* M *be a* $B \otimes_A C$*-module. Then*

$$\begin{aligned}
H_n(A, B, M) &= H_n(C, B \otimes_A C, M) \\
H^n(A, B, M) &= H^n(C, B \otimes_A C, M)
\end{aligned} \qquad \text{for } n = 0, 1, 2.$$

Proof Let R be a free DG resolution of the A-algebra B. Since R is a projective resolution of the A-module B, $H_i(R \otimes_A C) = \mathrm{Tor}_i^A(B, C) = 0$ for $i = 1, 2$, and $H_0(R \otimes_A C) = B \otimes_A C$. The DG C-algebra $R \otimes_A C$ is free over C. By the proof of Theorem 1.2.6, we can extend $R \otimes_A C$ to a free DG resolution S of the C-algebra $B \otimes_A C$ adjoining variables only in degrees ≥ 4; in particular $S_j = R_j \otimes_A C$ for $0 \leq j \leq 3$, and so

$$\begin{aligned}
H_n(C, B \otimes_A C, M) &= H_n(\Omega_{S|C} \otimes_S M) = H_n(\Omega_{R \otimes_A C|C} \otimes_{R \otimes_A C} M) \\
&= H_n(\Omega_{R|A} \otimes_R M) = H_n(A, B, M),
\end{aligned}$$

for $n = 0, 1, 2$, and similarly for cohomology. $\qquad\square$

Proposition 1.4.4 *Let A be a noetherian ring, B an A-algebra of finite type. Then we can choose a free DG resolution R of B so that $\Omega_{R|A} \otimes_R B$ is a complex of B-modules of finite type, and so if M is a B-module of finite type, $H_n(A, B, M)$ and $H^n(A, B, M)$ are B-modules of finite type.*

Proof Corollary 1.2.8. $\qquad\square$

Proposition 1.4.5 (Universal coefficient exact sequences) *Let $A \to B \to C$ be ring homomorphisms, M a C-module.*

 a) If N is a flat C-module,
$$H_n(A, B, M \otimes_C N) = H_n(A, B, M) \otimes_C N, \quad n = 0, 1, 2.$$

 b) If N is a flat C-module, A noetherian and B an A-algebra of finite type,
$$H^n(A, B, M \otimes_C N) = H^n(A, B, M) \otimes_C N, \quad n = 0, 1, 2.$$

 c) If M is an injective C-module,
$$H^n(A, B, M) = \operatorname{Hom}_C(H_n(A, B, C), M), \quad n = 0, 1, 2.$$

 d) There exist exact sequences
$$\operatorname{Tor}_2^C(\Omega_{B|A} \otimes_B C, M) \to H_1(A, B, C) \otimes_C M$$
$$\to H_1(A, B, M) \to \operatorname{Tor}_1^C(\Omega_{B|A} \otimes_B C, M) \to 0$$
$$0 \to \operatorname{Ext}_C^1(\Omega_{B|A} \otimes_B C, M) \to H^1(A, B, M)$$
$$\to \operatorname{Hom}_C(H_1(A, B, C), M) \to \operatorname{Ext}_C^2(\Omega_{B|A} \otimes_B C, M).$$

Proof Suppose N is a flat C-module, A noetherian and B an A-algebra of finite type. Choose a free DG resolution R of B so that $\Omega_{R|A} \otimes_R B$ is a complex of B-modules of finite type. Then $\operatorname{Hom}_B(\Omega_{R|A} \otimes_R B, M \otimes_C N) = \operatorname{Hom}_B(\Omega_{R|A} \otimes_R B, M) \otimes_C N$, and so $H^n(A, B, M \otimes_C N) = H^n(\operatorname{Hom}_B(\Omega_{R|A} \otimes_R B, M \otimes_C N)) = H^n(\operatorname{Hom}_B(\Omega_{R|A} \otimes_R B, M) \otimes_C N) = H^n(A, B, M) \otimes_C N$. This proves b); c) and a) are similar and d) follows from the universal coefficient spectral sequences
$$E_{p,q}^2 = \operatorname{Tor}_p^C(H_q(\Omega_{R|A} \otimes_R C), M) \Rightarrow H_{p+q}(\Omega_{R|A} \otimes_R M)$$
$$E_2^{p,q} = \operatorname{Ext}_C^p(H_q(\Omega_{R|A} \otimes_R C), M) \Rightarrow H^{p+q}(\operatorname{Hom}_C(\Omega_{R|A} \otimes_R C, M)),$$

where R is a free DG resolution of the A-algebra B (in the appendix we give a direct proof without using spectral sequences). $\qquad\square$

Proposition 1.4.6 (Jacobi–Zariski exact sequences) *Let $A \to B \to$*
C be ring homomorphisms and M a C-module. There exist exact se-
quences

$$H_2(A, B, M) \to H_2(A, C, M) \to H_2(B, C, M) \to$$
$$H_1(A, B, M) \to H_1(A, C, M) \to H_1(B, C, M) \to$$
$$H_0(A, B, M) \to H_0(A, C, M) \to H_0(B, C, M) \to 0$$

$$0 \to H^0(B, C, M) \to H^0(A, C, M) \to H^0(A, B, M) \to$$
$$H^1(B, C, M) \to H^1(A, C, M) \to H^1(A, B, M) \to$$
$$H^2(B, C, M) \to H^2(A, C, M) \to H^2(A, B, M).$$

Proof Let R be a free DG resolution of the A-algebra B and S a DG
A-algebra free over R (and in particular over A) such that it is a resolution of C (Theorem 1.2.6). Then $B \otimes_R S$ is a DG B-algebra free
over B. Consider the maps $S \to B \otimes_R S \to C$. The first map induces
an isomorphism in homology, since $R \to B$ induces an isomorphism in
homology and S is free over R, and also the composite $S \to C$ induces
an isomorphism in homology by the definition of S. So the second map
$B \otimes_R S \to C$ also induces an isomorphism in homology, i.e., $B \otimes_R S$
is a free DG resolution of the B-algebra C. Now the exact sequence of
complexes of free C-modules

$$0 \to (\Omega_{R|A} \otimes_R B)_B C \to \Omega_{S|A} \otimes_S C \to \Omega_{B \otimes_R S|B} \otimes_{B \otimes_R S} C \to 0$$

induces the required exact sequences.   □

Proposition 1.4.7 (Localization)

     a) Let A be a ring, T a multiplicative subset of A, B a $T^{-1}A$-algebra
     and M a B-module. Then

$$H_n(A, B, M) = H_n(T^{-1}A, B, M), \; n = 0, 1, 2$$
$$H^n(A, B, M) = H^n(T^{-1}A, B, M), \; n = 0, 1, 2.$$

     b) Let A be a ring, B an A-algebra, T a multiplicative subset of B
     and M a $T^{-1}B$-module. Then

$$H_n(A, B, M) = H_n(A, T^{-1}B, M), \; n = 0, 1, 2$$
$$H^n(A, B, M) = H^n(A, T^{-1}B, M), \; n = 0, 1, 2.$$

c) Let A be a ring, B an A-algebra, T a multiplicative subset of B and M a B-module. Then

$$T^{-1}H_n(A, B, M) = H_n(A, B, T^{-1}M), \ n = 0, 1, 2$$

and if A is noetherian and B an A-algebra of finite type, then also

$$T^{-1}H^n(A, B, M) = H^n(A, B, T^{-1}M), \ n = 0, 1, 2.$$

Proof a) follows from base change (1.4.3) by the homomorphism $A \to T^{-1}A$, and c) follows from 1.4.5, a) and b). For b), note first that the case $n = 0$ is clear by (1.4.1.b). Let $A' \to B$ be a surjective homomorphism of A-algebras where A' is a polynomial A-algebra. By (1.4.6) and (1.4.1) we have $H_2(A, B, M) = H_2(A', B, M)$ and an exact sequence

$$0 \to H_1(A, B, M) \to H_1(A', B, M) \to H_0(A, A', M)$$

and the same isomorphism and exact sequence with B replaced by $T^{-1}B$. So it is enough to prove the result for A' instead of A.

Let then U be the inverse image of T in A'. By a) and (1.4.3) we have

$$H_n(A', T^{-1}B, M) = H_n(U^{-1}A', T^{-1}B, M) =$$
$$H_n(U^{-1}A', B \otimes_A U^{-1}A, M) = H_n(A', B, M).$$

$\square$

Proposition 1.4.8 (Direct limits) *Let $\{A_i \to B_i\}_{i \in I}$ be a (filtered) direct system of ring homomorphisms and $\{M_i\}_{i \in I}$ a direct system of $\{B_i\}_{i \in I}$-modules. Then*

$$H_n(\varinjlim A_i, \varinjlim B_i, \varinjlim M_i) = \varinjlim H_n(A_i, B_i, M_i).$$

Proof For each i, construct a free DG resolution of the A_i-algebra B_i as follows. Choose R_i such that $(R_i)_0$ is the polynomial A_i-algebra over the elements of B_i, $(R_i)_0 = A_i[\{X_b\}_{b \in B_i}]$, $(R_i)_1$ the free $(R_i)_0$-module over the elements of $\ker((R_i)_0 \to B_i)$, and so on. Then there exists a direct system $\{R_i\}_{i \in I}$ with maps sending variables to variables such that $R = \varinjlim R_i$ is a free DG resolution of the $\varinjlim A_i$-algebra $\varinjlim B_i$, and then the result follows from the fact that the module of differentials of Definition 1.3.1 commutes with direct limits (details can be checked using Lemma 1.2.10 and its proof; alternatively, this same proof works with Definition 1.1.1 instead of Definition 1.3.6), and that direct limits are exact. $\square$

2

Formally smooth homomorphisms

In this chapter we prove some important results on smooth homomorphisms. Starting from basic definitions (Section 2.2), we interpret the first cohomology module in terms of infinitesimal extensions (Section 2.1) to characterize formal smoothness in terms of homology.

The first key result is the Jacobian criterion of formal smoothness (2.3.5), a homological characterization of this property. This result and some other results on the homology of field extensions in Section 2.4 (in particular a homological characterization of separability) will allow us to prove the main theorems in this section:

- Grothendieck's Theorems 2.5.8 and 2.5.9, which assert that formal smoothness over a field is equivalent to geometric regularity [EGA, 0_{IV}, 22.5.8]. Another ingenious proof by Faltings of this result can be seen in Matsumura's book [Mt, Theorem 28.7].
- Corollary 2.6.5, which reduces formal smoothness over a noetherian local ring to formal smoothness (or geometric regularity) over a field. This result was proved also by Grothendieck [EGA, 0_{IV} 19.7.1] and it is stated without proof in [Mt, Theorem 28.9]. The more difficult part is to prove that a formally smooth homomorphism is flat. Note that this is much easier if the homomorphism is of finite type (see, e.g., [Bo, Chap. X, §7.10, lemma 5], but the general case is needed.

Our proofs follow those by André [An1, 7.27, 16.18], but we need a few changes in order to avoid the use of simplicial methods or homology modules in higher dimensions. For example, Lemma 2.5.4 would be immediate from Jacobi–Zariski exact sequence and Corollary 2.5.3 if we could make use of the characterization of complete intersection rings by the vanishing of some H_3 [An1, 6.27].

Though these homological methods are spread along this book, it is perhaps in this section where they appear with more naturality, showing their great power in Commutative Algebra. Some suggestions for further reading on these methods could be [An1], [Qu], as well as many papers mainly by André himself, Avramov, Brezuleanu and Radu.

2.1 Infinitesimal extensions

Definition 2.1.1 Let A be a ring, B an A-algebra and M a B-module. An infinitesimal extension of B over A by M is an exact sequence of A-module homomorphisms

$$0 \to M \xrightarrow{i} E \xrightarrow{p} B \to 0,$$

where E is also an A-algebra, p an A-algebra homomorphism and $ei(x) = i(p(e)x)$ for all $e \in E$, $x \in M$. This last equality implies that M, as an ideal of E, is of square zero.

Conversely, if $p \colon E \to B$ is a surjective homomorphism of A-algebras whose kernel is an ideal M of square zero, then we have an infinitesimal extension of B over A by M.

We say that two infinitesimal extensions of B over A by M

$$0 \to M \to E \to B \to 0$$
$$0 \to M \to E' \to B \to 0$$

are equivalent if there exists an A-algebra homomorphism $f \colon E \to E'$ making the diagram

$$
\begin{array}{ccccccccc}
0 & \to & M & \to & E & \to & B & \to & 0 \\
  &   & \| &   & \downarrow f &   & \| &   &   \\
0 & \to & M & \to & E' & \to & B & \to & 0
\end{array}
$$

commute. In such a case, f is an isomorphism.

We denote by $\mathrm{Exalcom}_A(B, M)$ the set of equivalence classes of infinitesimal extensions of B over A by M.

We say that an infinitesimal extension

$$0 \to M \xrightarrow{i} E \xrightarrow{p} B \to 0$$

is trivial if there exists an A-algebra homomorphism $q \colon B \to E$ such that $pq = 1$.

Lemma 2.1.2 *Let A be a ring, B an A-algebra and M a B-module.*

Then there exists a trivial extension of B over A by M. Moreover, all trivial extensions of B over A by M are equivalent.

Proof Let $E = M \oplus B$ as A-module, with the A-algebra structure given by

$$(x, b)(x', b') = (bx' + b'x, bb'), x, x' \in M, b, b' \in B.$$

The exact sequence

$$0 \to M \xrightarrow{i} E \xrightarrow{p} B \to 0,$$

where $i(x) = (x, 0), p(x, b) = b$, is a trivial infinitesimal extension of B over A by M ($q \colon B \to E$ can be defined as $q(b) = (0, b)$).
   Now, let

$$0 \to M \xrightarrow{i'} E' \xrightarrow{p'} B \to 0$$

be a trivial infinitesimal extension of B over A by M. Let $q' \colon B \to E$ be such that $p'q' = 1$. Then the A-algebra homomorphism $f \colon E \to E'$, $f(x, b) = i'(x) + q'(b)$ makes the diagram

$$
\begin{array}{ccccccccc}
0 & \to & M & \xrightarrow{i} & E & \xrightarrow{p} & B & \to & 0 \\
& & \| & & \downarrow f & & \| & & \\
0 & \to & M & \xrightarrow{i'} & E' & \xrightarrow{p'} & B & \to & 0
\end{array}
$$

commute. $\qquad\qquad\qquad\qquad\qquad\qquad\qquad\qquad\qquad\qquad\qquad\square$

Definition 2.1.3 Let $f \colon C \to B$ be an A-algebra homomorphism, M a B-module. We define a map

$$f^* = \mathrm{Exalcom}_A(f, M) \colon \mathrm{Exalcom}_A(B, M) \to \mathrm{Exalcom}_A(C, M)$$

as follows: let

$$\mathbb{E} \colon 0 \to M \to E \to B \to 0$$

be an infinitesimal extension of B over A by M. The fibre product $E \times_B C$ gives a commutative diagram

$$
\begin{array}{ccccccccc}
\mathbb{E}^f \colon & 0 & \to & M & \to & E \times_B C & \to & C & \to & 0 \\
& & & \| & & \downarrow & & \downarrow f & & \\
\mathbb{E} \colon & 0 & \to & M & \to & E & \to & B & \to & 0
\end{array}
$$

such that $\mathbb{E}^f$ is an extension of C over A by M.
   If $\mathbb{E}$, $\mathbb{E}'$ are two equivalent extensions of B over A by M, then $\mathbb{E}^f$, $\mathbb{E}'^f$ are also equivalent. So we define $f^*([\mathbb{E}]) = [\mathbb{E}^f]$.

Theorem 2.1.4 *Let A be a ring, B an A-algebra and M a B-module. There exists a bijective map*

$$H^1(A, B, M) \to \mathrm{Exalcom}_A(B, M)$$

satisfying:

a) *the element $0 \in H^1(A, B, M)$ goes to the class of trivial extensions*

b) *if $f\colon C \to B$ is an A-algebra homomorphism, the canonical homomorphism*

$$H^1(A, B, M) \to H^1(A, C, M)$$

induced by f goes to

$$f^* = \mathrm{Exalcom}_A(f, M)\colon\ \mathrm{Exalcom}_A(B, M) \to \mathrm{Exalcom}_A(C, M).$$

Proof Let R be a polynomial ring over A and J an ideal of R such that $B = R/J$. We have an exact sequence (1.4.6), (1.4.1)

$$\mathrm{Der}_A(R, M) \xrightarrow{\alpha} \mathrm{Hom}_R(J, M) \to H^1(A, B, M) \to 0,$$

where for $d \in \mathrm{Der}_A(R, M)$, $\alpha(d) = d_{|J}$. So an element of $H^1(A, B, M)$ is represented by an R-module homomorphism $f\colon J \to M$ and two homomorphisms $f, g\colon J \to M$ represent the same element of $H^1(A, B, M)$ if and only if there exists an A-derivation $d\colon R \to M$ such that $f - g = d_{|J}$.

Let

$$\mathbb{E}\colon 0 \to M \to E \to B \to 0$$

be an infinitesimal extension of B over A by M. Since R is a polynomial ring over A, we have a commutative diagram

$$
\begin{array}{ccccccccc}
0 & \to & J & \to & R & \to & B & \to & 0 \\
 & & \downarrow f' & & \downarrow f & & \| & & \\
0 & \to & M & \to & E & \to & B & \to & 0.
\end{array}
$$

Moreover, if $g\colon R \to E$, $g'\colon J \to M$ are two other homomorphisms making the diagram commutative, then $d\colon R \to M$ defined by $d(x) = f(x) - g(x)$ is an A-derivation and $d_{|J} = f' - g'$. So we define a map $u\colon \mathrm{Exalcom}_A(B, M) \to H^1(A, B, M)$ by $u([\mathbb{E}]) = [f']$.

Conversely, let $f\colon J \to M$ be an R-module homomorphism. We have a commutative diagram

$$
\begin{array}{ccccccccc}
 & & 0 & \to & J & \to & R & \xrightarrow{p} & B & \to & 0 \\
 & & & & \downarrow f & & \downarrow & & \| & & \\
\mathbb{E}\colon & & 0 & \to & M & \to & E_f & \to & B & \to & 0,
\end{array}
$$

where $E_f = M \oplus_J R$ is the quotient of the A-algebra $M \oplus R$ (with multiplication $(x,r)(x',r') = (rx' + r'x, rr')$, $x, x' \in M, r, r' \in R$, as in the proof of Lemma 2.1.2, where we are considering M as an R-module via p) by the ideal generated by the elements $\{(f(s), -s) : s \in J\}$. We have then an extension $\mathbb{E}$ of B over A by M. If $f, g : J \to M$ are two R-module homomorphisms such that there is some $d \in \mathrm{Der}_A(R, M)$ such that $f - g = d_{|J}$, then the associated extensions are equivalent. In effect, the A-algebra homomorphism $h : E_f \to E_g$ defined by $h(x, r) = (x + d(r), r)$ is well defined and makes the diagram

$$
\begin{array}{ccccccccc}
0 & \to & M & \to & E_f & \to & B & \to & 0 \\
 & & \| & & \downarrow & & \| & & \\
0 & \to & M & \to & E_g & \to & B & \to & 0
\end{array}
$$

commute.

We then define $v : H^1(A, B, M) \to \mathrm{Exalcom}_A(B, M)$ by $v([f]) = [\mathbb{E}]$.

It is easy to check that u and v are inverse, and properties a) and b). $\square$

2.2 Formally smooth algebras

Definition 2.2.1 Let A be a ring, B an A-algebra and J an ideal of B. We say that the A-algebra B is formally smooth in the J-adic topology (or J-smooth) if the following holds:

for any surjective A-algebra homomorphism $p : E \to C$ with kernel an ideal of square zero, and for any A-algebra homomorphism $f : B \to C$ that is continuous for the J-adic topology in B and the discrete topology in C (i.e., such that $f(J^n) = 0$ for some n), there exists an A-algebra homomorphism $g : B \to E$ such that $pg = f$:

$$
\begin{array}{c}
B \\
{\scriptstyle g}\swarrow \quad \downarrow{\scriptstyle f} \\
0 \to N \to E \xrightarrow{\ p\ } C \to 0.
\end{array}
$$

Note that then g is continuous for the discrete topology in E, since $f(J^n) = 0 \implies g(J^{2n}) = 0$.

In the case $J = 0$, we say that B is a smooth A-algebra. So smooth implies J-smooth for any ideal J.

Examples 2.2.2

a) A polynomial ring over A is a smooth A-algebra.

b) If A is a ring and S a multiplicative subset of A, then $S^{-1}A$ is a smooth A-algebra, since if N is a square zero ideal of a ring E and $e \in E$ is such that $e + N$ is a unit in E/N, then e is a unit in E (let $x \in E$ be such that $ex - 1 \in N$. Since $N^2 = 0$, we have $0 = (ex - 1)^2 = e^2 x^2 - 2ex + 1$ and so $1 = e(2x - ex^2)$).

c) Let $A \to B$, $f \colon B \to C$ be ring homomorphisms, J an ideal of B and T an ideal of C such that $f(J) \subset T$. If B is a J-smooth A-algebra and C is a T-smooth B-algebra, then C is a T-smooth A-algebra.

Lemma 2.2.3 *Let $A \to B$ be a ring homomorphism, J an ideal of B, C an A-algebra and T an ideal of C. Assume that B is a J-smooth A-algebra and C complete for the T-adic topology. Let $f_1 \colon B \to C/T$ be a continuous A-algebra homomorphism (i.e., $f_1(J^n) = 0$ for some n). Then there exists a continuous A-algebra homomorphism $f \colon B \to C$ inducing f_1.*

Proof Since B is J-smooth, there exists a continuous A-algebra homomorphism $f_2 \colon B \to C/T^2$ making the diagram

$$
\begin{array}{ccccccccc}
 & & & & & B & & & \\
 & & & {}^{f_2}\!\swarrow & & \downarrow {\scriptstyle f_1} & & & \\
0 & \to & T/T^2 & \to & C/T^2 & \longrightarrow & C/T & \to & 0
\end{array}
$$

commute. Similarly, if we have already constructed $f_n \colon B \to C/T^n$, by the J-smoothness of B there exists $f_{n+1} \colon B \to C/T^{n+1}$ inducing f_n. We define $f = \varprojlim f_n \colon B \to C = \varprojlim C/T^n$. $\square$

Lemma 2.2.4 *Let A be a ring, B an A-algebra, J an ideal of B and $0 \le i \le 2$ an integer. The following are equivalent:*

a) $\varinjlim H^i(A, B/J^n, M) = 0$ *for any B-module M annihilated by J.*

b) $\varinjlim H^i(A, B/J^n, M) = 0$ *for any B-module M annihilated by some power of J (here the limit is taken over the direct system for n large enough).*

Proof b) implies a) is trivial. To prove the converse, let M be a B-module such that $J^t M = 0$ for some t. We will see by induction on

t that $\varinjlim H^i(A, B/J^n, M) = 0$. The case $t = 1$ holds by assumption. Assume that the equality is valid for $t - 1$. The exact sequence

$$0 \to JM \to M \to M/JM \to 0$$

induces an exact sequence

$$\varinjlim H^i(A, B/J^n, JM) \to \varinjlim H^i(A, B/J^n, M)$$
$$\to \varinjlim H^i(A, B/J^n, M/JM).$$

Since JM is annihilated by J^{t-1}, the term on the left is zero. The one on the right is also zero, since M/JM is annihilated by J. So $\varinjlim H^i(A, B/J^n, M) = 0$. □

Lemma 2.2.5 *Let A be a ring, B an A-algebra, J an ideal of B. The following are equivalent:*

   a) B is an A-algebra formally smooth for the J-adic topology
   b) $\varinjlim H^1(A, B/J^n, M) = 0$ for any B/J-module M.

Proof b) $\implies$ a) Let $p \colon E \to C$ be a surjective A-algebra homomorphism with kernel N satisfying $N^2 = 0$. Let $\mathbb{E}$ be the extension

$$0 \to N \to E \to C \to 0.$$

Let $f \colon B \to C$ be an A-algebra homomorphism such that $f(J^t) = 0$. Then f induces $f_t \colon B/J^t \to C$, and N is a B/J^t-module. Consider the extension $\mathbb{E}^{f_t}$ (2.1.3). We have a commutative diagram

$$
\begin{array}{ccccccccc}
0 & \to & N & \to & E^{f_t} & \xrightarrow{p_t} & B/J^t & \to & 0 \\
& & \| & & \downarrow g & & \downarrow f_t & & \\
0 & \to & N & \xrightarrow{i} & E & \xrightarrow{p} & C & \to & 0
\end{array}
$$

where $E^{f_t} = E \times_C B/J^t$.
   By (2.1.4) and (2.2.4),

$$\varinjlim \mathrm{Exalcom}_A(B/J^n, N) = \varinjlim H^1(A, B/J^n, N) = 0.$$

Then there exists an integer $s \geq t$ such that the extension $\mathbb{E}^{f_s} = (\mathbb{E}^{f_t})^{q_{st}}$ is trivial, where $q_{st} \colon B/J^s \to B/J^t$ is the canonical map:

$$
\begin{array}{ccccccccc}
\mathbb{E}^{f_s} \colon & 0 & \to & N & \to & E^{f_s} & \xrightarrow{p_s} & B/J^s & \to & 0 \\
& & & \| & & \downarrow \theta & & \downarrow q_{st} & & \\
\mathbb{E}^{f_t} \colon & 0 & \to & N & \xrightarrow{i} & E^{f_t} & \xrightarrow{p_t} & B/J^t & \to & 0,
\end{array}
$$

with $E^{f_s} = E^{f_t} \times_{B/J^t} B/J^s$. That is, there exists an A-algebra homomorphism $\sigma \colon B/J^s \to E^{f_s}$ such that $p_s\sigma = 1$. Therefore $p_t\theta\sigma = q_{st}$.

Let $\pi_s \colon B \to B/J^s$ be the canonical map

$$
\begin{array}{ccc}
 & B & \\
 & \Big\downarrow{\scriptstyle \pi_s} & \\
{\scriptstyle \theta\sigma} \nearrow & B/J^s & \\
 & \Big\downarrow{\scriptstyle q_{st}} & \\
E^{f_t} \;\xrightarrow{\;p_t\;} & B/J^t & \\
\Big\downarrow{\scriptstyle g} & \Big\downarrow{\scriptstyle f_t} & \\
E \;\xrightarrow{\;p\;} & C. &
\end{array}
$$

We have $pg\theta\sigma\pi_s = f_t p_t \theta\sigma\pi_s = f_t q_{st}\pi_s = f$. So B is a J-smooth A-algebra.

a) $\implies$ b) Let

$$0 \to M \to E \xrightarrow{\;\varepsilon_n\;} B/J^n \to 0$$

be an infinitesimal extension of B/J^n over A by M. Let $\pi_n \colon B \to B/J^n$ be the canonical map. Since B is formally smooth over A, there exists $\psi_n \colon B \to E$ such that $\pi_n = \varepsilon_n\psi_n$. Since $M^2 = 0$, we deduce that $\psi_n(J^{2n}) = 0$ and so there exists $\phi_n \colon B/J^{2n} \to E$ such that $\psi_n = \phi_n\pi_{2n}$. Let E^q be as above:

$$
\begin{array}{ccccccc}
0 & \to & M & \longrightarrow & E^q & \longrightarrow & B/J^{2n} & \to & 0 \\
 & & \| & & \downarrow & {\scriptstyle \phi_n}\nearrow \;\; \downarrow{\scriptstyle q} & & \\
0 & \to & M & \longrightarrow & E & \longrightarrow & B/J^n & \to & 0.
\end{array}
$$

The homomorphism ϕ_n induces an A-algebra homomorphism $B/J^{2n} \to E^q$, which is a section of the map $E^q \to B/J^{2n}$. Therefore,

$$0 \to M \to E^q \to B/J^{2n} \to 0$$

is a trivial extension. By (2.1.4) $\varinjlim H^1(A, B/J^n, M) = 0$. $\qquad\square$

2.3 Jacobian criteria

Theorem 2.3.1 (Jacobian criterion for smoothness) *Let A be a ring and B an A-algebra. The following are equivalent:*

a) B is a smooth A-algebra

b) $H^1(A, B, M) = 0$ for any B-module M

c) $H_1(A, B, B) = 0$ and $\Omega_{B|A}$ is a projective B-module

d) For any presentation $B = R/I$ where R is a polynomial ring over A, the B-module homomorphism

$$I/I^2 \to \Omega_{R|A} \otimes_R B$$

has left inverse.

Proof a) $\iff$ b) is a consequence of (2.2.5) for $J = 0$, b) $\iff$ c) is a consequence of the universal coefficient exact sequence (1.4.5.d), and c) $\iff$ d) is a consequence of the exact sequence (1.4.1), (1.4.6)

$$0 \to H_1(A, B, B) \to I/I^2 \to \Omega_{R|A} \otimes_R B \to \Omega_{B|A} \to 0.$$

$\square$

Next we study the nondiscrete case.

Lemma 2.3.2 *Let A be a ring, I an ideal of A, $B = A/I$ and M a B-module. We have a natural injective B-module homomorphism*

$$H^2(A, B, M) \to \operatorname{Ext}_A^1(I, M).$$

Proof In Definition 1.1.1, take $R = A$, F a free A-module with an exact sequence of A-modules

$$0 \to U \to F \to I \to 0.$$

Let U_0 be as in (1.1.1). We have $H^2(A, B, M) =$

$$\operatorname{coker}(\operatorname{Hom}_B(F/IF, M) = \operatorname{Hom}_A(F, M) \to \operatorname{Hom}_B(U/U_0, M)$$
$$= \operatorname{Hom}_A(U/U_0, M)).$$

The result follows from the commutative diagram of exact rows and columns

$$
\begin{array}{ccccccc}
& & 0 & & & & \\
& & \downarrow & & & & \\
\operatorname{Hom}_A(F, M) & \to & \operatorname{Hom}_A(U/U_0, M) & \to & H^2(A, B, M) & \to & 0 \\
\| & & \downarrow & & \downarrow & & \\
\operatorname{Hom}_A(F, M) & \to & \operatorname{Hom}_A(U, M) & \to & \operatorname{Ext}_A^1(I, M) & \to & 0.
\end{array}
$$

$\square$

Lemma 2.3.3 *Let A be a noetherian ring, I an ideal of A, $n > 0$ an integer. Then there exists an integer $s \geq n$ such that the canonical homomorphism*

$$\operatorname{Tor}_1^A(I^s, A/I) \to \operatorname{Tor}_1^A(I^n, A/I)$$

is zero.

Proof We have $\operatorname{Tor}_1^A(I^n, A/I) = \operatorname{Tor}_2^A(A/I^n, A/I) = \operatorname{Tor}_1^A(A/I^n, I)$. Let

$$0 \to U \to F \to I \to 0$$

be an exact sequence of A-modules with F free of finite type. From the exact sequence

$$0 \to \operatorname{Tor}_1^A(A/I^n, I) \to U/I^n U \to F/I^n F$$

we obtain $\operatorname{Tor}_1^A(A/I^n, I) = (U \cap I^n F)/I^n U$. Artin–Rees lemma [Mt, Theorem 8.5] says that there exists an integer $r > 0$ such that $I^t F \cap U = I^{t-r}(I^r F \cap U)$ for all $t > r$. Let $s = n + r$. Then $I^s F \cap U \subset I^n U$, and so the map

$$\operatorname{Tor}_1^A(A/I^s, I) \to \operatorname{Tor}_1^A(A/I^n, I)$$

is zero. $\qquad\square$

Lemma 2.3.4 *Let A be a ring, I an ideal of A, M an A/I-module. Then*

$$\varinjlim H^1(A, A/I^n, M) = 0.$$

If A is noetherian, we also have

$$\varinjlim H^2(A, A/I^n, M) = 0.$$

Proof By (1.4.1), $\varinjlim H^1(A, A/I^n, M) = \varinjlim \operatorname{Hom}_{A/I^n}(I^n/I^{2n}, M) = \varinjlim \operatorname{Hom}_{A/I}(I^n/I^{n+1}, M)$, and this last module is zero, since the map $I^{n+1}/I^{n+2} \to I^n/I^{n+1}$ is zero. This proves the first part.

Now assume A is noetherian. Consider the exact sequence

$$0 \to \operatorname{Ext}_{A/I}^1(I^n/I^{n+1}, M) \to \operatorname{Ext}_A^1(I^n, M)$$
$$\to \operatorname{Hom}_{A/I}(\operatorname{Tor}_1^A(I^n, A/I), M),$$

which comes from the change-of-rings spectral sequence (see the Appendix for a direct proof without spectral sequences)

$$E_2^{p,q} = \mathrm{Ext}_{A/I}^p(\mathrm{Tor}_q^A(I^n, A/I), M) \Rightarrow \mathrm{Ext}_A^{p+q}(I^n, M).$$

Since $\varinjlim \mathrm{Ext}_{A/I}^1(I^n/I^{n+1}, M) = 0$ and $\varinjlim \mathrm{Hom}_{A/I}(\mathrm{Tor}_1^A(I^n, A/I), M) = 0$ by (2.3.3), we deduce that $\varinjlim \mathrm{Ext}_A^1(I^n, M) = 0$.

On the other hand, by (2.3.2) we have an injective homomorphism

$$\varinjlim H^2(A, A/I^n, M) \to \varinjlim \mathrm{Ext}_A^1(I^n, M)$$

which shows that $\varinjlim H^2(A, A/I^n, M) = 0$.   □

Theorem 2.3.5 (Jacobian criterion of formal smoothness) *Let A be a ring, B a noetherian A-algebra and J an ideal of B. The following are equivalent:*

    a) B is a formally smooth A-algebra for the J-adic topology.
    b) $H^1(A, B, M) = 0$ for any B/J-module M.
    c) $H_1(A, B, B/J) = 0$ and $\Omega_{B|A} \otimes_B B/J$ is a projective B/J-module.
    d) For any presentation $B = R/I$ where R is a smooth A-algebra, the B/J-module homomorphism

$$I/I^2 \otimes_B B/J \to \Omega_{R|A} \otimes_R B/J$$

    has left inverse.

    (The implications a)$\Longleftarrow$b) $\Longleftrightarrow$ c) $\Longleftrightarrow$ d) hold without the noetherian assumption.)

Proof For any B/J-module M we have the exact sequence (1.4.6)

$$H^1(B, B/J^n, M) \to H^1(A, B/J^n, M) \to H^1(A, B, M) \to$$
$$H^2(B, B/J^n, M),$$

hence an exact sequence

$$\varinjlim H^1(B, B/J^n, M) \to \varinjlim H^1(A, B/J^n, M) \to H^1(A, B, M) \to$$
$$\varinjlim H^2(B, B/J^n, M).$$

    a) $\Longrightarrow$ b) By (2.3.4) $\varinjlim H^1(A, B/J^n, M) = H^1(A, B, M)$ and the result follows from (2.2.5).

    b) $\Longrightarrow$ a) By (2.3.4) $\varinjlim H^1(A, B/J^n, M) \to H^1(A, B, M)$ is injective and so the result follows again from (2.2.5).

    b) $\Longleftrightarrow$ c) $\Longleftrightarrow$ d) follows as in the proof of (2.3.1 b) $\Longleftrightarrow$ c) $\Longleftrightarrow$ d)).
□

We now give an example which shows that the noetherian assumption is essential. This is a negative answer to a question of Brezuleanu [Br, Remark 1.3.(i)].

Lemma 2.3.6 *Let A be a ring, N an ideal of A such that $N^2 = N$. Let $I \subset N$ be another ideal of A such that $I \neq IN$. Let $B = A/I$, $J = N/I$. Then the A-algebra B is formally smooth for the J-adic topology and there exists a B/J-module M such that $H^1(A, B, M) \neq 0$.*

Proof Since $J = J^2$, for any $B/J = A/N$-module M we have

$$\varinjlim H^1(A, B/J^n, M) = H^1(A, B/J, M) = H^1(A, A/N, M)$$
$$= \mathrm{Hom}_{A/N}(N/N^2, M) = 0$$

since $N^2 = N$. So B is formally smooth over A by (2.2.5).

On the other hand,

$$H^1(A, B, M) = H^1(A, A/I, M) = \mathrm{Hom}_{A/I}(I/I^2, M)$$
$$= \mathrm{Hom}_{A/N}(I/I^2 \otimes_{A/I} A/N, M) = \mathrm{Hom}_{A/N}(I/IN, M),$$

which is not zero in general (take $M = I/IN$), since $I \neq IN$. □

Example 2.3.7 We show that there exists a ring A satisfying the assumptions of Lemma 2.3.6. Let $A = C(\mathbb{R}, \mathbb{R})$ be the ring of continuous functions from the real field $\mathbb{R}$ on itself. Let $i \in C(\mathbb{R}, \mathbb{R})$ be the identity map, and let $I = (i)$ be the ideal of A generated by i. We know that i is not a unit in A since $i(0) = 0$, and is not a zero divisor since if $\phi i = 0$ for $\phi \in A$, then $\phi(x) = 0$ for all $x \in \mathbb{R}$, $x \neq 0$, and so $\phi = 0$ by continuity. The ideal I is not absolutely isolated [Bo, Chapitre II, §2, Ex. 15, c], and so $\mathrm{rad}(I) \neq I$ [Bo, loc. cit., a]. In particular, I is not a prime ideal.

Now let N be a maximal ideal of A such that $I \subset N$. N is prime and so $N^2 = N$ [Bo, loc. cit. d]. Finally, we see that $I \neq IN$. If $I = IN$, $i \in IN = iN$, so $i = i\psi$ for some $\psi \in N$. Then $i(1 - \psi) = 0$. Since i is not a zero divisor, $\psi = 1$, but this is impossible since $\psi \in N$.

Remark 2.3.8 Let A be a ring, $(B, \mathfrak{n}, L)$ a noetherian local A-algebra essentially of finite type, $J \subset \mathfrak{n}$ an ideal of B. Then B is a smooth A-algebra if and only if it is J-smooth if and only if it is $\mathfrak{n}$-smooth. By definition, it is enough to show that $\mathfrak{n}$-smooth implies smooth. By (2.3.5), $\mathfrak{n}$-smooth implies $H_1(A, B, L) = 0$. By (1.4.5.d) we have then

$\operatorname{Tor}_1^B(\Omega_{B|A}, L) = 0$, and since B is noetherian and $\Omega_{B|A}$ a B-module of finite type, we deduce that $\Omega_{B|A}$ is a projective B-module. Therefore, by (1.4.5.d) again, $H_1(A, B, B) \otimes_B L = 0$, and so $H_1(A, B, B) = 0$ by Nakayama's lemma and (1.4.4), (1.4.7). By (2.3.1) B is a smooth A-algebra.

2.4 Field extensions

Let $F|K$ be a field extension, M an F-module. By (1.4.5) we have

$$H_n(K, F, M) = H_n(K, F, F) \otimes_F M$$
$$H^n(K, F, M) = \operatorname{Hom}_F(H_n(K, F, F), M)$$

for $n = 0, 1, 2$, so we concentrate on the F-vector spaces $H_n(K, F, F)$.

Proposition 2.4.1 *If $F|K$ is a field extension, then $H_2(K, F, F) = 0$.*

Proof By (1.4.8) we can assume that $F|K$ is of finite type, and then by (1.4.6) we can assume $F = K(t)$. If t is transcendental over K, then F is isomorphic to the field of fractions of the polynomial ring $K[X]$ and so $H_2(K, F, F) = 0$ by (1.4.1) and (1.4.7). If t is algebraic over K, then $F = K[X]/(f)$ with $f \in K[X]$. Since f is not a zero divisor, we can take $R = K[X]$, $I = (f)$, $F = K[X]$, $U = 0$ in Definition 1.1.1, and therefore $H_2(K, F, F) = 0$. $\square$

Corollary 2.4.2 *Let $F|K$, $E|F$ be field extensions. There exists an exact sequence*

$$0 \to H_1(K, F, E) \to H_1(K, E, E) \to H_1(F, E, E) \to$$
$$\Omega_{F|K} \otimes_F E \to \Omega_{E|K} \to \Omega_{E|F} \to 0.$$

Proof This follows from (1.4.1), (1.4.6), (2.4.1). $\square$

Proposition 2.4.3 *Let $F|K$ be a field extension with $F = K(t)$. Then:*

   a) *In the case that t is transcendental over K, $H_1(K, F, F) = 0$ and $\dim_F \Omega_{F|K} = 1$.*
   b) *If t is separable algebraic over K, then $H_1(K, F, F) = 0$ and $\Omega_{F|K} = 0$.*
   c) *If t is algebraic but inseparable over K, then $\dim_F H_1(K, F, F) = 1$ and $\dim_F \Omega_{F|K} = 1$.*

Proof a) $F = K(t)$ is isomorphic to the field of fractions of $K[X]$. So $H_1(K, F, F) = 0$ by (1.4.1) and (1.4.7). Also $\Omega_{F|K} = \Omega_{K[X]|K} \otimes_{K[X]} F$, and so $\dim_F \Omega_{F|K} = 1$.

b) and c) We have a surjective K-algebra homomorphism $K[X] \to F$ with kernel (f), where f is the irreducible polynomial of t over K. Applying (1.4.6) to the sequence

$$K \to K[X] \to F$$

we obtain an exact sequence

$$0 \to H_1(K, F, F) \to (f)/(f^2) \overset{d}{\to} \Omega_{K[X]|K} \otimes_{K[X]} F \to \Omega_{F|K} \to 0.$$

Since $\dim_F(f)/(f^2) = 1$ and $\dim_F \Omega_{K[X]|K} \otimes_{K[X]} F = 1$, we have $\dim_F H_1(K, F, F) = \dim_F \Omega_{F|K} \leq 1$. So we have to show that t is separable if and only if $\dim_F H_1(K, F, F) = 0$.

We have that t is separable over K if and only if $f'(t) \neq 0$. We have isomorphisms

$$
\begin{array}{ccccc}
\Omega_{K[X]|K} \otimes_{K[X]} F & \overset{\approx}{\longrightarrow} & K[X] \otimes_{K[X]} F & \overset{\approx}{\longrightarrow} & F \\
g dX \otimes 1 & \mapsto & g \otimes 1 & \mapsto & g(t)
\end{array}
$$

and so an exact sequence

$$0 \to H_1(K, F, F) \to (f)/(f^2) \overset{d}{\longrightarrow} F$$

with $d(gf + (f^2)) = g(t)f'(t)$.

Then t is separable if and only if $f'(t) \neq 0$ and this is equivalent to $d \neq 0$. Since $\dim_F(f)/(f^2) = 1$, this is also equivalent to the injectivity of d, and this last condition is equivalent to $H_1(K, F, F) = 0$. $\qquad\square$

Theorem 2.4.4 (Cartier equality) *Let $F|K$ be a field extension of finite type. We have*

$$\operatorname{tr} \deg(F|K) = \dim_F \Omega_{F|K} - \dim_F H_1(K, F, F).$$

Proof By (2.4.2) we may assume that $F = K(t)$, and so the result follows from (2.4.3). $\qquad\square$

Theorem 2.4.5 *A general field extension $F|K$ is separable if and only if $H_1(K, F, F) = 0$.*

Proof Assume $F|K$ separable. By (1.4.8) we can assume that $F|K$ is of finite type and separably generated. By (1.4.6), we can reduce the problem to the case $F = K(t_1, \ldots, t_n)$ where $t_1, \ldots, t_n$ are algebraically independent over K, and to the case of a finite separable extension $F|K$. In either case, using (1.4.6) once more, we reduce the problem to the case $F = K(t)$. Then (2.4.3) gives $H_1(K, F, F) = 0$.

Conversely, assume now that $H_1(K, F, F) = 0$. We must show that any finitely generated subextension $L|K$ of $F|K$ is separably generated. Applying (2.4.2) to $K \to L \to F$ we see that $H_1(K, F, F) = 0$ implies $H_1(K, L, L) = 0$. So we may assume that $F|K$ is finitely generated. By (2.4.4), $\operatorname{tr deg} F|K = \dim_F \Omega_{F|K}$, that is, if $n = \operatorname{tr deg}(F|K)$, there exist n elements $(x_1, \ldots, x_n)$ such that $\{dx_1, \ldots, dx_n\}$ is an F-basis of $\Omega_{F|K}$. Let $E = K(x_1, \ldots, x_n)$. In the exact sequence

$$\Omega_{E|K} \otimes_E F \to \Omega_{F|K} \to \Omega_{F|E} \to 0$$

the map on the left is surjective, and so $\Omega_{F|E} = 0$. By (2.4.4) again, we see that $F|E$ is algebraic and $H_1(E, F, F) = 0$. Then $\operatorname{tr deg} E|K = \operatorname{tr deg} F|K = n$, so $x_1, \ldots, x_n$ are algebraically independent over K, and so they form a transcendence basis of $F|K$. We must see that $F|E$ is separable. Let $t \in F$. Applying (2.4.2) to $E \to E(t) \to F$ and bearing in mind that $H_1(E, F, F) = 0$, we see that $H_1(E, E(t), E(t)) = 0$. By (2.4.3) t is then separable over E. $\qquad\square$

Corollary 2.4.6 *A field extension $F|K$ is separable if and only if F is a smooth K-algebra.*

Proof (2.3.1). $\qquad\square$

Now we shall prove some lemmas on algebras over a field which we use in the next sections.

Lemma 2.4.7 *Let A be a ring containing a field of characteristic $p > 0$. Let B be a reduced A-algebra and M a B-module. Then the canonical homomorphism*

$$H_1(A^p, B^p, M) \xrightarrow{\alpha} H_1(A, B, M)$$

is zero (here A^p is the subring of A of elements a^p with $a \in A$, and similarly B^p).

Proof Let R be a polynomial A-algebra with a surjective A-algebra homomorphism $R \to B$. We have a commutative diagram

$$
\begin{array}{ccc}
H_1(A^p, B^p, M) & \xrightarrow{\alpha} & H_1(A, B, M) \\
\downarrow & & \downarrow \gamma \\
H_1(R^p, B^p, M) & \xrightarrow{\beta} & H_1(R, B, M).
\end{array}
$$

By (1.4.1) $H_1(A, R, M) = 0$ and so by (1.4.6) γ is injective. So, $\alpha = 0$ if $\beta = 0$. Thus we may assume $B = A/I$ with I an ideal of A. Since B is reduced, $B^p = A^p/J$, with $J = \{x^p : x \in I\}$. Then α is the map

$$
J/J^2 \otimes_{B^p} M = H_1(A^p, B^p, M) \xrightarrow{\alpha} H_1(A, B, M) = I/I^2 \otimes_B M,
$$

which is zero since $J/J^2 \to I/I^2$ is zero. $\qquad \square$

Lemma 2.4.8 *Let K be a field of characteristic $p > 0$, $F|K$ a field extension such that $K^{1/p} \subset F$. Then the canonical homomorphism*

$$
H_1(K, F, M) \to H_1(K^{1/p}, F, M)
$$

is zero for any F-module M.

Proof By (1.4.5) we may assume that M is an algebraically closed field L. In the commutative diagram

$$
\begin{array}{ccc}
H_1(K, F, L) & \xrightarrow{\alpha} & H_1(K^{1/p}, F, L) \\
\downarrow & & \downarrow \gamma \\
H_1(K, L, L) & \xrightarrow{\beta} & H_1(K^{1/p}, L, L)
\end{array}
$$

$\beta = 0$ by (2.4.7) and γ is injective by (2.4.2). So $\alpha = 0$. $\qquad \square$

Lemma 2.4.9 *Let K be a field of characteristic $p > 0$, $(B, \mathfrak{n}, L)$ a local K-algebra and F a field such that $K \subset F \subset K^{1/p}$. Then the ring $B \otimes_K F$ is local with maximal ideal*

$$
J = \{z \in B \otimes_K F : z^p \in \mathfrak{n} \otimes_K F\}.
$$

Proof From the exact sequence $0 \to \mathfrak{n} \to B \to L \to 0$ we obtain an exact sequence

$$
0 \to \mathfrak{n} \otimes_K F \to B \otimes_K F \to L \otimes_K F \to 0.
$$

Clearly J is a proper ideal of $B \otimes_K F$. So it is enough to show that if $x \in B \otimes_K F - J$, then x is a unit in $B \otimes_K F$. The element x^p belongs to the subring $B \otimes_K K$ of $B \otimes_K F$, since $F^p \subset K$. Moreover $x^p \notin \mathfrak{n} \otimes_K F$

since $x \notin J$. So we can identify x^p to an element of $B - \mathfrak{n}$. Therefore x^p is a unit, and so is x. □

Lemma 2.4.10 *Let K be a field of characteristic $p > 0$, B a local K-algebra, F the residue field of the local ring $B \otimes_K K^{1/p}$ (Lemma 2.4.9) and M an F-module. There exists a natural isomorphism*

$$H_2(B \otimes_K K^{1/p}, F, M) = H_1(K, B, M).$$

Proof In the commutative diagram

$$
\begin{array}{ccc}
H_1(K, B, M) & \longrightarrow & H_1(K, F, M) \\
\downarrow{\alpha} & & \downarrow{\gamma} \\
H_1(K^{1/p}, B \otimes_K K^{1/p}, M) & \xrightarrow{\beta} & H_1(K^{1/p}, F, M)
\end{array}
$$

$\gamma = 0$ by (2.4.8) and α is an isomorphism by (1.4.3). So $\beta = 0$. By the Jacobi–Zariski exact sequence associated to $K^{1/p} \to B \otimes_K K^{1/p} \to F$ and (2.4.1)

$$0 \to H_2(B \otimes_K K^{1/p}, F, M)$$
$$\xrightarrow{\delta} H_1(K^{1/p}, B \otimes_K K^{1/p}, M) \xrightarrow{\beta} H_1(K^{1/p}, F, M)$$

we deduce that δ is an isomorphism. The required isomorphism is then $\alpha^{-1}\delta$. □

Lemma 2.4.11 *Let $A \to F \to L$ be ring homomorphisms such that F and L are fields, and A contains a field. Then the canonical homomorphism*

$$H_2(A, F, L) \xrightarrow{h} H_2(A, L, L)$$

is an isomorphism.

Proof By (1.4.6) and (2.4.1) h is surjective. Let K be a field contained in A. Applying (1.4.6) to the sequences $K \to A \to L$ and $K \to A \to F$ we obtain a commutative diagram with exact rows

$$
\begin{array}{ccccc}
0 = H_2(K, F, L) & \to & H_2(A, F, L) & \to & H_1(K, A, L) \\
 & & \downarrow{h} & & \| \\
0 = H_2(K, L, L) & \to & H_2(A, L, L) & \to & H_1(K, A, L)
\end{array}
$$

(2.4.1), (1.4.5). So h is injective. □

2.5 Geometric regularity

Lemma 2.5.1 *Let $(A, \mathfrak{m}, K)$ be a noetherian local ring, I an ideal of A, $a_1, \ldots, a_n$ a set of generators of I, $F = A^n$ and $j \colon F \to I$ the surjective homomorphism that sends the canonical basis of F to $a_1, \ldots, a_n$. Let $H_1(a_1, \ldots, a_n; A)$ be the first Koszul homology module associated to the set of generators $a_1, \ldots, a_n$ of I. We have an exact sequence*

$$0 \to H_2(A, A/I, M) \to H_1(a_1, \ldots, a_n; A) \otimes_{A/I} M \to$$

$$F/IF \otimes_{A/I} M \xrightarrow{\bar{j}} I/I^2 \otimes_{A/I} M \to 0$$

for any A/I-module M, where $\bar{j}$ is the homomorphism induced by j.

Proof This follows easily from Definition 1.1.1, bearing in mind that, with the notation of (1.1.1), $U/U_0 = H_1(a_1, \ldots, a_n; A)$. $\qquad\square$

Theorem 2.5.2 *Let $(A, \mathfrak{m}, K)$ be a noetherian local ring and I an ideal of A. The following are equivalent:*

  a) I is generated by a regular sequence
  b) $H_2(A, A/I, K) = 0$
  c) $H_2(A, A/I, -) = 0$.

Proof a) $\implies$ c) Let $a_1, \ldots, a_n$ be a regular sequence that generates the ideal I and let F, j and $H_1(a_1, \ldots, a_n; A)$ be as in (2.5.1). Then $H_1(a_1, \ldots, a_n; A) = 0$ and so $H_2(A, A/I, -) = 0$ by (2.5.1).

b) $\implies$ a) Let $a_1, \ldots, a_n$ be a minimal set of generators of the ideal I. Let F, j and $H_1(a_1, \ldots, a_n; A)$ be as in (2.5.1). In the exact sequence of (2.5.1)

$$0 \to H_2(A, A/I, K) \to H_1(a_1, \ldots, a_n; A) \otimes_{A/I} K \to F/IF \otimes_{A/I} K$$

$$\xrightarrow{\bar{j}} I/I^2 \otimes_{A/I} K \to 0$$

we have $H_2(A, A/I, K) = 0$ and $\bar{j}$ is an isomorphism, since it can be identified with the map $F/\mathfrak{m}F \to I/\mathfrak{m}I$ induced by j. Therefore $H_1(a_1, \ldots, a_n; A) \otimes_{A/I} K = 0$, so $H_1(a_1, \ldots, a_n; A) = 0$ by Nakayama's Lemma, and $a_1, \ldots, a_n$ is a regular sequence. $\qquad\square$

Corollary 2.5.3 *Let $(A, \mathfrak{m}, K)$ be a noetherian local ring. The following are equivalent:*

  a) A is a regular local ring

b) $H_2(A, K, K) = 0$. $\qquad\qquad\qquad\qquad\qquad\qquad\qquad\qquad\qquad\square$

Lemma 2.5.4 *Let* $K \to (B, \mathfrak{n}, L)$ *be a local homomorphism of regular local rings. Then*

$$H_2(K, B, L) = 0.$$

Proof Let $b_1, \ldots, b_n$ be a regular sequence that generates the ideal $\mathfrak{n}$. Let $f\colon K[X_1, \ldots, X_n] \to B$ be the K-algebra homomorphism sending X_i to b_i. In the Jacobi–Zariski exact sequence associated to $K \to K[X_1, \ldots, X_n] \xrightarrow{f} B$

$$H_2(K, K[X_1, \ldots, X_n], L) \to H_2(K, B, L) \to H_2(K[X_1, \ldots, X_n], B, L)$$
$$\to H_1(K, K[X_1, \ldots, X_n], L)$$

the first and last terms are zero by (1.4.1). Thus we have to show that $H_2(K[X_1, \ldots, X_n], B, L) = 0$.

Let $g\colon K[X_1, \ldots, X_n] \to K$ be the K-algebra homomorphism sending X_i to 0. Since $X_1, \ldots, X_n$ is a regular sequence in $K[X_1, \ldots, X_n]$, the Koszul complex $\mathbb{K}(X_1, \ldots, X_n; K[X_1, \ldots, X_n])$ is a free resolution of the $K[X_1, \ldots, X_n]$-module $K[X_1, \ldots, X_n]/(X_1, \ldots, X_n) = K$. Then, considering K as a $K[X_1, \ldots, X_n]$-module via g, we have

$$\mathrm{Tor}_j^{K[X_1, \ldots, X_n]}(K, B) = H_j(X_1, \ldots, X_n; B) = H_j(b_1, \ldots, b_n; B) = 0$$

for all $j > 0$.

Then by (1.4.3),

$$H_2(K[X_1, \ldots, X_n], B, L) = H_2(K, B \otimes_{K[X_1, \ldots, X_n]} K, L)$$
$$= H_2(K, B/(b_1, \ldots, b_n), L) = H_2(K, L, L)$$

and this last term vanishes by (2.4.1), (2.5.3) and the Jacobi–Zariski exact sequence associated to K, its residue field, and L. $\qquad\square$

Remark Using André–Quillen H_3 we can show that if K is a regular local ring, the vanishing of $H_2(K, B, L)$ is equivalent to B being a complete intersection.

Lemma 2.5.5 *Let* K *be a field,* $(B, \mathfrak{n}, L)$ *a noetherian local K-algebra such that* $H_1(K, B, L) = 0$. *Then* B *is a regular local ring.*

Proof Applying (1.4.6) and (2.4.1) to $K \to B \to L$

$$0 = H_2(K, L, L) \to H_2(B, L, L) \to H_1(K, B, L) = 0$$

we obtain $H_2(B, L, L) = 0$ and so the result follows from (2.5.3). □

Definition 2.5.6 Let K be a field, B a noetherian K-algebra. We say that B is geometrically regular over K if for any finite field extension $F|K$, the noetherian ring $B \otimes_K F$ is regular.

The next three theorems show in particular that a noetherian local algebra over a field is geometrically regular if and only if it is formally smooth (for the adic topology of its maximal ideal).

Theorem 2.5.7 *Let K be a field, $(B, \mathfrak{n}, L)$ a noetherian local K-algebra. If B is formally smooth for the $\mathfrak{n}$-adic topology, then B is geometrically regular over K.*

Proof Let $F|K$ be a finite field extension, $C = B \otimes_K F$, N a maximal ideal of C. We have to show that C_N is a regular local ring. C is a finite B-algebra and so integral over B. In particular $\mathfrak{n}C \subset N$, and so any C/N-module is an L-module. Let M be a C/N-module. By (1.4.3), $H^1(F, C, M) = H^1(K, B, M)$, and this last term is zero by (2.3.5). Again by (2.3.5), C is formally smooth over F for the N-adic topology. Since C_N is a smooth C-algebra (2.2.2.b) we obtain that C_N is a formally smooth F-algebra (2.2.2.c). Then, by (2.3.5) and (2.5.5), C_N is a regular local ring. □

Theorem 2.5.8 *Let K be a field, $(B, \mathfrak{n}, L)$ a noetherian local K-algebra such that the extension $L|K$ is separable. The following are equivalent:*

 a) B is a formally smooth K-algebra for the $\mathfrak{n}$-adic topology
 b) B is a geometrically regular K-algebra
 c) B is a regular local ring.

Proof a) $\Longrightarrow$ b) is (2.5.7). b) $\Longrightarrow$ c) is clear. To see c) $\Longrightarrow$ a), consider the Jacobi–Zariski exact sequence

$$H_2(B, L, L) \to H_1(K, B, L) \to H_1(K, L, L).$$

The right term is zero by (2.4.5), and the left one is also zero by (2.5.3). So $H_1(K, B, L) = 0$ and then B is formally smooth by (2.3.5). □

Theorem 2.5.9 *Let K be a field of characteristic $p > 0$, $(B, \mathfrak{n}, L)$ a noetherian local K-algebra. The following are equivalent:*

a) *B is a formally smooth K-algebra for the $\mathfrak{n}$-adic topology.*

b) *B is a geometrically regular K-algebra.*

c) *For any finite field extension $F|K$ such that $F \subset K^{1/p}$, the noetherian local ring (2.4.9) $B \otimes_K F$ is regular.*

d) *The local ring $B \otimes_K K^{1/p}$ is regular (and in particular noetherian).*

Proof a) $\Longrightarrow$ b) is (2.5.7). b) $\Longrightarrow$ c) is clear. To see c) $\Longrightarrow$ d), let $K^{1/p} = \varinjlim F_i$, with $F_i|K$ finite extensions such that $F_i \subset K^{1/p}$. Let E_i be the residue field of $B \otimes_K F_i$, and E the residue field of $B \otimes_K K^{1/p}$. By assumption, each local ring $B \otimes_K F_i$ is regular and so by (2.5.3), (1.4.5), $H_2(B \otimes_K F_i, E_i, E) = 0$. By (2.4.11), $H_2(B \otimes_K F_i, E_i, E) = H_2(B \otimes_K F_i, E, E)$, and so by (1.4.8),

$$0 = \varinjlim H_2(B \otimes_K F_i, E_i, E) = \varinjlim H_2(B \otimes_K F_i, E, E)$$
$$= H_2(\varinjlim B \otimes_K F_i, E, E) = H_2(B \otimes_K K^{1/p}, E, E).$$

So by (2.5.3) it is enough to show that $B \otimes_K K^{1/p}$ is noetherian. Let Q be the field of fractions of B, and $B^{1/p} = \{x \in Q^{1/p} : x^p \in B\}$. We have a ring isomorphism $B^{1/p} \to B$ given by $b \mapsto b^p$. Since B is regular, we deduce that $B^{1/p}$ is regular. Let $F|K$ be a finite field extension such that $F \subset K^{1/p}$. We have a canonical homomorphism

$$\alpha_F \colon B \otimes_K F \to B^{1/p}, \alpha_F(b \otimes x) = bx,$$

which is local by (2.4.9). This homomorphism is integral, since if $y \in B^{1/p}$, $y^p = \alpha_F(y^p \otimes 1)$, and so $\dim(B \otimes_K F/\ker(\alpha_F)) = \dim(B^{1/p})$. The extension $F|K$ is finite, so $\dim B = \dim(B \otimes_K F)$, and so $\dim(B^{1/p}) = \dim(B \otimes_K F)$. Therefore α_F is injective since $B \otimes_K F$, being a regular local ring, is a domain. Let $n = \dim(B \otimes_K F)$ and $\mathfrak{p} = (x_1, \ldots, x_n)$ its maximal ideal. By base change, the homomorphism $(B \otimes_K F)/\mathfrak{p} \to B^{1/p}/\alpha_F(\mathfrak{p})B^{1/p}$ induced by α_F is an integral extension and therefore $\dim(B^{1/p}/\alpha_F(\mathfrak{p})B^{1/p}) = \dim((B \otimes_K F)/\mathfrak{p}) = 0$.

Hence $\alpha_F(\mathfrak{p})B^{1/p} = \big(\alpha_F(x_1), \ldots, \alpha_F(x_n)\big)$ is an ideal of height n, and so $\alpha_F(x_1), \ldots, \alpha_F(x_n)$ is a regular sequence in $B^{1/p}$ [Mt, Theorem 17.4]. Thus $\mathrm{gr}_{\mathfrak{p}}(B \otimes_K F) = ((B \otimes_K F)/\mathfrak{p})[X_1, \ldots, X_n]$, and

$$\mathrm{gr}_{\mathfrak{p}}(B^{1/p}) = \mathrm{gr}_{\alpha_F(\mathfrak{p})B^{1/p}}(B^{1/p}) = (B^{1/p}/\alpha_F(\mathfrak{p})B^{1/p})[X_1, \ldots, X_n]$$

[Mt, Theorem 16.2.i]. By the local flatness criterion [Mt, Theorem 22.3],

α_F is flat. Then $\varinjlim \alpha_{F_i} : \varinjlim B \otimes_K F_i = B \otimes_K K^{1/p} \to B^{1/p}$ is flat and so faithfully flat, since it is local. Since $B^{1/p}$ is noetherian, so is $B \otimes_K K^{1/p}$.

d) $\implies$ a) We have $0 = H_2(B \otimes_K K^{1/p}, E, E) = H_1(K, B, E) = H_1(K, B, L) \otimes_L E$ by (2.5.3), (2.4.10), and (1.4.5) respectively. Then B is a formally smooth K-algebra for the $\mathfrak{n}$-adic topology by (2.3.5). $\square$

2.6 Formally smooth local homomorphisms of noetherian rings

Proposition 2.6.1 *Let A be a ring, B and C two A-algebras such that the canonical homomorphism $A \to B$ is surjective. Then for any $B \otimes_A C$-module M there exists a natural exact sequence*

$$H_2(A, B, M) \to H_2(C, B \otimes_A C, M) \to \operatorname{Tor}_1^A(B, C) \otimes_C M$$
$$\to H_1(A, B, M) \to H_1(C, B \otimes_A C, M) \to 0.$$

Proof Let $I = \ker(A \to B)$, and suppose that F is a free A-module such that there exists an exact sequence

$$0 \to U \to F \xrightarrow{j} I \to 0$$

of A-modules.

Let ϕ, U_0 be as in (1.1.1). We have a commutative diagram with exact rows

$$0 \to \operatorname{Tor}_2^A(B, M) = \operatorname{Tor}_1^A(I, M) \to U \otimes_A M \quad \to F \otimes_A M \to I \otimes_A M$$

$$\quad\quad \downarrow{e} \quad\quad\quad\quad \downarrow{p} \quad\quad \| \quad\quad \|$$

$$0 \to H_2(A, B, M) \quad\quad \to U/U_0 \otimes_A M \to F \otimes_A M \to I \otimes_A M$$

from which we deduce $\ker e = \ker p$.

The surjective homomorphism $\phi \otimes 1 : \bigwedge_A^2 F \otimes_A M \to U_0 \otimes_A M$ and the exact sequence

$$U_0 \otimes_A M \xrightarrow{v} U \otimes_A M \xrightarrow{p} U/U_0 \otimes_A M$$

give isomorphisms $\ker e = \ker p = \operatorname{im} v = \operatorname{im}(v(\phi \otimes 1))$. So we have an exact sequence

$$\bigwedge_A^2 F \otimes_A M \to \operatorname{Tor}_2^A(B, M) \xrightarrow{e} H_2(A, B, M).$$

On the other hand, we have an exact sequence

$$0 \to IC \to C \to B \otimes_A C \to 0$$

and a surjective C-module homomorphism $F \otimes_A C \to IC$, where $F \otimes_A C$ is a free C-module, and so the above reasoning shows that there exists an exact sequence

$$\bigwedge_C^2 (F \otimes_A C) \otimes_C M$$
$$\|$$
$$\bigwedge_A^2 F \otimes_A M \quad \to \quad \mathrm{Tor}_2^C(B \otimes_A C, M) \quad \to \quad H_2(C, B \otimes_A C, M).$$

Consider the exact sequence

$$\mathrm{Tor}_2^A(B, M) \to \mathrm{Tor}_2^C(B \otimes_A C, M) \to$$
$$\mathrm{Tor}_1^A(B, C) \otimes_C M \to \mathrm{Tor}_1^A(B, M) \to \mathrm{Tor}_1^C(B \otimes_A C, M) \to 0$$

associated to the change-of-rings spectral sequence (see the Appendix for a direct proof without using spectral sequences)

$$E_{p,q}^2 = \mathrm{Tor}_p^C(\mathrm{Tor}_q^A(B, C), M) \Rightarrow \mathrm{Tor}_{p+q}^A(B, M).$$

Then the required exact sequence follows from the diagram

$$
\begin{array}{ccc}
\bigwedge_A^2 F \otimes_A M & = & \bigwedge_A^2 F \otimes_A M \\
\downarrow & & \downarrow \\
\mathrm{Tor}_2^A(B, M) \to & \mathrm{Tor}_2^C(B \otimes_A C, M) & \to \mathrm{Tor}_1^A(B, C) \otimes_C M \to \\
\downarrow & & \downarrow \qquad \nearrow \\
H_2(A, B, M) \to & H_2(C, B \otimes_A C, M) & \\
\downarrow & & \downarrow \\
0 & & 0
\end{array}
$$

$$\to \mathrm{Tor}_1^A(B, M) \to \mathrm{Tor}_1^C(B \otimes_A C, M) \to 0$$

and from the isomorphisms of 1.4.1.c, $H_1(A, B, M) = \mathrm{Tor}_1^A(B, M)$ and $H_1(C, B \otimes_A C, M) = \mathrm{Tor}_1^C(B \otimes_A C, M)$. $\qquad\square$

Lemma 2.6.2 *Let A be a ring, B and C two A-algebras and M a $B \otimes_A C$-module. Then the canonical homomorphism*

$$H_1(A, B, M) \to H_1(C, B \otimes_A C, M)$$

is surjective.

Proof Let R be a polynomial A-algebra such that $B = R/I$. We have an exact sequence

$$I \otimes_A C \to R \otimes_A C \xrightarrow{p} B \otimes_A C \to 0.$$

Let $J = \ker p$. We have an epimorphism $I \otimes_A C \to J$. The Jacobi–Zariski exact sequences associated to $A \to R \to B$ and $C \to R_C \to B_C$, where $R_C = R \otimes_A C$ and $B_C = B \otimes_A C$, give a commutative diagram with exact rows

$$
\begin{array}{ccccccc}
0 \to & H_1(A, B, M) & \to & I \otimes_R M & \to & \Omega_{R|A} \otimes_R M & \to & \Omega_{B|A} \otimes_B M \\
& \downarrow & & \downarrow & & \| & & \| \\
0 \to & H_1(C, B_C, M) & \to & J \otimes_{R_C} M & \to & \Omega_{R_C|C} \otimes_{R_C} M & \to & \Omega_{B_C|C} \otimes_{B_C} M.
\end{array}
$$

Then it is enough to show that $I \otimes_R M \to J \otimes_{R \otimes_A C} M$ is surjective. But $I \otimes_R M = (I \otimes_A C) \otimes_{R \otimes_A C} M$, and the epimorphism $I \otimes_A C \to J$ gives an epimorphism $(I \otimes_A C) \otimes_{R \otimes_A C} M \to J \otimes_{R \otimes_A C} M$. □

Lemma 2.6.3 *Let* $(A, \mathfrak{m}, K) \to (B, \mathfrak{n}, L)$ *be a local homomorphism of noetherian local rings, such that* $H_1(A, B, L) = 0$. *Then*

    a) $H_2(A, K, L) \to H_2(B, B \otimes_A K, L)$ *is surjective*
    b) $H_1(A, K, L) \to H_1(B, B \otimes_A K, L)$ *is injective.*

Proof By (2.6.2), since $H_1(A, B, L) = 0$ we deduce $H_1(K, B \otimes_A K, L) = 0$, and so $H_2(K, B \otimes_A K, L) = 0$ by (2.5.5), (2.5.4). We have Jacobi–Zariski exact sequences associated to $A \to K \to B \otimes_A K$ and $A \to B \to B \otimes_A K$

$$\cdots \to H_n(A, K, L) \xrightarrow{\alpha_n} H_n(A, B \otimes_A K, L) \to H_n(K, B \otimes_A K, L) \to \cdots$$

$$\cdots \to H_n(A, B, L) \to H_n(A, B \otimes_A K, L) \xrightarrow{\beta_n} H_n(B, B \otimes_A K, L) \to \cdots$$

The homomorphisms of a) and b) are $\beta_2 \alpha_2$ and $\beta_1 \alpha_1$ respectively. Since α_1 and β_1 are injective, and α_2 and β_2 are surjective, we obtain the desired result. □

Theorem 2.6.4 *Let* $(A, \mathfrak{m}, K) \to (B, \mathfrak{n}, L)$ *be a local homomorphism of noetherian local rings. The following are equivalent:*

    a) $H_1(A, B, L) = 0$
    b) $H_1(K, B \otimes_A K, L) = 0$ *and* B *is a flat* A-module.
        Moreover, if these conditions hold, then $H_2(A, B, L) = 0$.

Proof a) $\Longrightarrow$ b) Consider the exact sequence of (2.6.1)

$$H_2(A, K, L) \to H_2(B, B \otimes_A K, L) \to \operatorname{Tor}_1^A(B, K) \otimes_B L \to H_1(A, K, L)$$
$$\to H_1(B, B \otimes_A K, L) \to 0.$$

By (2.6.3) $\operatorname{Tor}_1^A(B, K) \otimes_B L = 0$. Since $\operatorname{Tor}_1^A(B, K)$ is a B-module of finite type, by Nakayama's lemma we have $\operatorname{Tor}_1^A(B, K) = 0$, and by the local flatness criterion [Mt, Theorem 22.3] B is a flat A-module. Finally, since $H_1(A, B, L) = 0$, by (1.4.3) we obtain $H_1(K, B \otimes_A K, L) = 0$.

b) $\Longrightarrow$ a) By (1.4.3), $H_1(A, B, L) = H_1(K, B \otimes_A K, L) = 0$.

Under these assumptions, $H_1(K, B \otimes_A K, L) = 0$ implies that also $H_2(K, B \otimes_A K, L) = 0$ by (2.5.5), (2.5.4) and then by (1.4.3),

$$H_2(A, B, L) = H_2(K, B \otimes_A K, L) = 0. \qquad \square$$

Corollary 2.6.5 *Let* $(A, \mathfrak{m}, K) \to (B, \mathfrak{n}, L)$ *be a local homomorphism of noetherian local rings. The following are equivalent:*

    a) B is a formally smooth A-algebra for the $\mathfrak{n}$-adic topology

    b) B is a flat A-module and the K-algebra $B \otimes_A K$ is geometrically regular.

Proof (2.6.4), (2.3.5), (2.5.8) and (2.5.9). $\qquad \square$

2.7 Appendix: The Mac Lane separability criterion

As a by-product of the theory so far, we now deduce Mac Lane's criterion for separability of a field extension.

Proposition 2.7.1 *Let* $L|K$ *be a field extension of characteristic* $p > 0$. *Then* $L|K$ *is separable if and only if* $L \otimes_K K^{1/p}$ *is a field.*

Proof This follows from (2.5.9), (a) $\Longleftrightarrow$ (d), (2.4.6) and (2.4.9). $\qquad \square$

3

Structure of complete noetherian local rings

In this chapter we prove Cohen's 1946 structure theorems for complete noetherian local rings (Theorem 3.2.4). Our proof follows Grothendieck [EGA, 0_{IV} §19.8] (and in some parts Bourbaki [Bo, Chapter IX]), thus using the results on formal smoothness proved in Section 2.

We would like to point out that this way can be reversed. One first can prove Cohen's theorems as in some standard books on Commutative Algebra (e.g., in Matsumura's [Mt, sections 28 and 29], treating first the equicharacteristic case), and then Radu [Ra2] shows how to use them to prove that a formally smooth homomorphism is flat (the more difficult part of (2.6.5)).

3.1 Cohen rings

Definition 3.1.1 Let $f \colon (A, \mathfrak{m}, K) \to (B, \mathfrak{n}, L)$ be a local homomorphism of noetherian local rings. We say that B is a Cohen A-algebra if B is complete, A-flat and $f(\mathfrak{m})B = \mathfrak{n}$.

Proposition 3.1.2 *Let* $(A, \mathfrak{m}, K) \to (B, \mathfrak{n}, L)$ *be a local homomorphism of noetherian local rings such that B is a Cohen A-algebra. If $L|K$ is separable then B is a formally smooth A-algebra (for the $\mathfrak{n}$-adic topology).*

Proof By (2.6.5) it is enough to show that the K-algebra $B \otimes_A K = L$ is geometrically regular, and this is equivalent to the separability of $L|K$ (2.4.6), (2.5.8), (2.5.9). □

Lemma 3.1.3 *Let A be a ring, I an ideal of A, and M an A-module which is Hausdorff for the I-adic topology. If $M/IM = 0$ then $M = 0$.*
□

Lemma 3.1.4 *Let A be a ring, I an ideal of A, M an A-module which is complete for the I-adic topology and N an A-module which is Hausdorff for the I-adic topology. Let $f\colon M \to N$ be an A-module homomorphism. If the induced homomorphism $\overline{f}\colon M/IM \to N/IN$ is surjective then f is surjective. If in addition N is A-flat and $\overline{f}$ is injective, then f is injective.*

Proof Let $K = \ker f$, $P = \operatorname{im}(f)$, $C = \operatorname{coker}(f)$. Since M is complete, we have that P is complete. So from the commutative diagram with exact rows and columns

$$
\begin{array}{ccccccccc}
 & & & & 0 & & & & \\
 & & & & \downarrow & & & & \\
0 & \to & P & \to & N & \to & C & \to & 0 \\
 & & \downarrow & & \downarrow & & \downarrow g & & \\
0 & \to & \widehat{P} & \to & \widehat{N} & \to & \widehat{C} & & \\
 & & \downarrow & & & & & & \\
 & & 0 & & & & & &
\end{array}
$$

we deduce that g is injective, i.e., C is Hausdorff. Since $\overline{f}$ is surjective, $C/IC = 0$ and so $C = 0$ by (3.1.3). Then we have an exact sequence $0 \to K \to M \xrightarrow{f} N \to 0$, that induces an exact sequence

$$0 \to K/IK \to M/IM \xrightarrow{\overline{f}} N/IN \to 0$$

when N is A-flat. Thus $K/IK = 0$. Since K is Hausdorff, being a submodule of a Hausdorff module, again by (3.1.3) we deduce $K = 0$. □

Lemma 3.1.5 *Let $\{(A_i, \mathfrak{m}_i), f_{ji}\colon A_i \to A_j\}_{i,j \in I}$ be a filtered direct system of local rings and flat local homomorphisms. Assume that $f_{ji}(\mathfrak{m}_i)A_j = \mathfrak{m}_j$ for all $i \leq j$. Let $A = \varinjlim A_i$, $\mathfrak{m} = \varinjlim \mathfrak{m}_i$, and for all $i \in I$, let $f_i\colon A_i \to A$ be the canonical homomorphism. Then*

    a) A is a local ring with maximal ideal $\mathfrak{m}$, and for all $i \in I$, the homomorphism $f_i\colon A_i \to A$ is local, flat and satisfies $f_i(\mathfrak{m}_i)A = \mathfrak{m}$.

    b) If $\mathfrak{m}_i$ is an A_i-module of finite type for some i, then the (Hausdorff) completion $\widehat{A}$ of the local ring A is a noetherian ring. If moreover A_i is noetherian for all i, then A is a noetherian ring.

Proof a) It is clear that $\mathfrak{m}$ is a proper ideal of A. If $x \in A - \mathfrak{m}$, there exist $i \in I$ and $x_i \in A_i$ such that $f_i(x_i) = x$. We have $x_i \notin \mathfrak{m}_i$, and so x_i is a unit in A_i, and then x is a unit in A. We deduce that A is a local ring with maximal ideal $\mathfrak{m}$. The equalities $f_i(\mathfrak{m}_i)A = \mathfrak{m}$ are clear. Finally, f_i is flat since $\varinjlim$ is exact and commutes with tensor products.

b) Assume $\mathfrak{m}_i$ is an ideal of finite type of A_i for some i. Then $\mathfrak{m} = f_i(\mathfrak{m}_i)A$ is an ideal of finite type of A. Thus $\widehat{\mathfrak{m}} = \mathfrak{m}\widehat{A}$ [AM, Proposition 10.13] and it is an ideal of finite type of $\widehat{A}$. Then $\widehat{A}$ is a noetherian ring ($\mathrm{gr}_{\widehat{\mathfrak{m}}}(\widehat{A})$ is an $\widehat{A}/\widehat{\mathfrak{m}}$-algebra generated by $\widehat{\mathfrak{m}}/\widehat{\mathfrak{m}}^2$, and so noetherian, now apply [AM, corollary 10.25], or alternatively [ZS, VIII, §3, Theorem 7, Corollary 4]).

Assume now that A_i is noetherian for all i. For all $i \in I$ and all $n > 0$,

$$A \otimes_{A_i} (A_i/\mathfrak{m}_i^n) = A/\mathfrak{m}^n = \widehat{A}/\widehat{\mathfrak{m}}^n = \widehat{A}/f_i(\mathfrak{m}_i^n)\widehat{A}.$$

Since A is flat as A_i-module, we deduce that $\widehat{A}/f_i(\mathfrak{m}_i^n)\widehat{A}$ is flat as $A_i/\mathfrak{m}_i^n$-module. Since $A_i \to \widehat{A}$ is a local homomorphism of noetherian rings (easy or see e.g. [Mt, Theorem 8.2] or [Bo, III, §2.13, Proposition 19]) the A_i-module $\widehat{A}$ is $\mathfrak{m}_i$-adically ideal-Hausdorff, and by the local flatness criterion [Mt, Theorem 22.3] $\widehat{A}$ is a flat A_i-module. Taking $\varinjlim$ we see that $\widehat{A}$ is a flat A-module, and therefore faithfully flat, since $A \to \widehat{A}$ is local. Then, since $\widehat{A}$ is noetherian, so is A. $\qquad\square$

Remarks

i) Note that if we do not assume $f_{ji}(\mathfrak{m}_i)A_j = \mathfrak{m}_j$ for $i \leq j$, but a) holds, then the proof shows that b) also holds (alternatively, note that a) implies $f_{ji}(\mathfrak{m}_i)A_j = \mathfrak{m}_j$ for $i \leq j$).

ii) A similar result avoiding the flatness assumption can be seen in [Og1].

Let $(A, \mathfrak{m}, K)$ be a local ring, $f \in A[X]$ a monic polynomial of degree $n > 0$. Let $B = A[X]/(f)$, which as A-module is free of rank n, with basis $\{1, x, \ldots, x^{n-1}\}$, where x is the image of X in B. Let $\overline{B} = B/\mathfrak{m}B = A[X]/(\mathfrak{m}A[X] + (f))$, and $\overline{f}$ the image of f in $K[X]$. Then $\overline{B} = K[X]/(\overline{f})$. Let $\overline{f} = \prod_i \overline{g_i}^{e_i}$ be the decomposition of $\overline{f}$ into irreducible factors in $K[X]$, with distinct monic polynomials $\overline{g_i}$. For each i, let $g_i \in A[X]$ be a representant of $\overline{g_i}$.

Lemma 3.1.6 *The ideals* $\mathfrak{m}_i = (\mathfrak{m} + g_i)B$ *of B are maximal and distinct. Moreover* $B/\mathfrak{m}_i = K[X]/(\overline{g_i})$. *There are no other maximal ideals of B.*

Proof Let $\bar{\mathfrak{m}}_i = (\bar{g}_i)\bar{B}$. The contraction of $\bar{\mathfrak{m}}_i$ from $\bar{B}$ to B is $\mathfrak{m}_i$. Since $\bar{\mathfrak{m}}_i$ is maximal ($\bar{B}/\bar{\mathfrak{m}}_i = K[X]/(\bar{g}_i)$ is a field) and $\bar{B}$ is a quotient of B, we see that $\mathfrak{m}_i$ is maximal. It is clear that the $\mathfrak{m}_i$ are distinct, and $B/\mathfrak{m}_i = K[X]/(\bar{g}_i)$. Let $\mathfrak{n}$ be a maximal ideal of B. Since B is an A-module of finite type, by Nakayama's lemma we have $\mathfrak{n} + \mathfrak{m}B \neq B$, and so $\mathfrak{m}B \subset \mathfrak{n}$. Then $\mathfrak{n}$ is the contraction in B of some maximal ideal of $\bar{B}$, i.e., some $(\bar{g}_i)\bar{B} = \bar{\mathfrak{m}}_i$, and so $\mathfrak{n} = \mathfrak{m}_i$. $\qquad\square$

Lemma 3.1.7 *Let $(A, \mathfrak{m}, K)$ be a noetherian local ring, $L = K(x)$ a field extension of K. Then there exist a noetherian local ring $(B, \mathfrak{n}, L)$ and a flat local homomorphism $f : (A, \mathfrak{m}, K) \to (B, \mathfrak{n}, L)$ such that $f(\mathfrak{m})B = \mathfrak{n}$.*

Proof First assume that x is algebraic over K. Let $\bar{f} \in K[X]$ be the irreducible polynomial of x over K, and let $f \in A[X]$ be a monic polynomial representing $\bar{f}$. Let $B = A[X]/(f)$. By (3.1.6) B is a noetherian local ring with maximal ideal $\mathfrak{n} := \mathfrak{m}B$, and B is a flat A-module (since it is free). The residue field of B is $K[X]/(\bar{f}) = L$.

Now assume x transcendental over K. Let $\mathfrak{q} = \mathfrak{m}A[X] = \mathfrak{m}[X]$ be the prime ideal extension of $\mathfrak{m}$ in $A[X]$. Let $B = A[X]_\mathfrak{q}$. Then B is a noetherian local ring with maximal ideal $\mathfrak{m}B$, and residue field $K(X) = L$. It is clear that B is flat as A-module. $\qquad\square$

Lemma 3.1.8 *Let $(A, \mathfrak{m}, K)$ be a noetherian local ring, $L|K$ a field extension. Then there exist a Cohen A-algebra $(B, \mathfrak{n}, L)$. If $L|K$ is separable then it is unique up to A-isomorphism.*

Proof Existence: let $\{F_i\}_{i \in I}$ be a family of subfields of L containing K, over a well-ordered indexing set I that possesses a greatest element ω, satisfying:

a) $i \leq j \implies F_i \subset F_j$;
b) the fields K and L belong to this family, K corresponds to the first element α of I, and $F_\omega = L$;
c) if $j \in I$ has a predecessor i, then the extension $F_j|F_i$ is generated by one element;
d) if $j \in I$ has no predecessor, then $F_j = \bigcup_{i<j} F_i$.

Such a family exists: for instance give L a well-ordering and define F_j as the K-subfield of L generated by the elements $i \in L$ such that $i < j$. Then the family $\{F_i\} \cup \{L\}$ works.

For each $j \in I$, we define a local ring $(C_j, \mathfrak{m}_j, F_j)$ and a homomorphism $f_j \colon A \to C_j$ by transfinite induction as follows: if $j = \alpha$ is the first element of I, then $C_j = A$. If j has a predecessor i, we define $f_{ji} \colon (C_i, \mathfrak{m}_i, F_i) \to (C_j, \mathfrak{m}_j, F_j)$ as in (3.1.7) and $f_j = f_{ji} f_i$. If j has no predecessor, $C_j = \varinjlim_{i<j} C_i$ and $f_j = \varinjlim_{i<j} f_i$.

Let I' be the set of $j \in I$ such that for all $i < j$ the homomorphism $f_i \colon A \to C_i$ is local flat and $\mathfrak{m}C_i = \mathfrak{m}_i$. If $I' \neq I$, let β be the smallest element of $I \setminus I'$. It is clear that the first element α of I is in I', and so $\emptyset \neq S_\beta := \{i \in I : i < \beta\} \subset I'$. If S_β does not have a maximum, then $C_\beta = \varinjlim_{i \in S_\beta} C_i$, and so $\beta \in I'$ by (3.1.5), which is a contradiction. If S_β has a maximum γ, we construct C_β from C_γ as in (3.1.7), and so we have again the contradiction $\beta \in I'$. Therefore $I' = I$, and so the homomorphism

$$f_\omega \colon (A, \mathfrak{m}, K) \to (C_\omega, \mathfrak{m}_\omega, F_\omega = L)$$

is local flat and $\mathfrak{m}C_\omega = \mathfrak{m}_\omega$.

Similarly we see that C_ω is noetherian: we consider the set I'' of elements $j \in I$ such that C_j is noetherian. Using (3.1.5) and (3.1.7) as before we see that $I'' = I$ and so C_ω is noetherian.

Thus we define $(B, \mathfrak{n}, L) := (\widehat{C}_\omega, \widehat{\mathfrak{m}}_\omega, L)$.

Uniqueness: now assume that $L|K$ is separable. Let $(B', \mathfrak{n}', L)$ be another Cohen A-algebra. By (3.1.2), B is $\mathfrak{n}$-smooth over A. By Lemma 2.2.3, the canonical composite homomorphism $B \to B/\mathfrak{n} = L \xrightarrow{=} B'/\mathfrak{n}' = L$ is induced by a continuous (i.e., local) homomorphism $f \colon B \to B'$. By (3.1.4) f is an isomorphism. $\qquad\square$

Definition 3.1.9 Let A be a local ring, and let $p \geq 0$ be the characteristic of its residue field. Let $P = \mathbb{Z}_{(p)}$ (localization of the ring of integers $\mathbb{Z}$ at the prime ideal $(p) = p\mathbb{Z}$). We say that A is a Cohen ring if the canonical homomorphism $P \to A$ makes A a Cohen A-algebra. For example, the completion of $\mathbb{Z}_{(p)}$ (i.e., the ring of p-adic integers) is a Cohen ring.

Proposition 3.1.10 *Let A be a local ring, and $p \geq 0$ the characteristic of its residue field. Then*

  a) If $p = 0$, then A is a Cohen ring if and only if it is a field (of characteristic zero).

b) *If $p \neq 0$, then A is a Cohen ring if and only if it is a complete discrete valuation domain with maximal ideal generated by $p \cdot 1$.*

Proof If A is a Cohen ring with $p > 0$, then by definition, the maximal ideal of A is generated by $p \cdot 1$. Since A is P-flat, $p \cdot 1$ is not a zero divisor in A and so A is a regular local ring of Krull dimension 1, i.e., a discrete valuation domain, and complete by definition. The case of $p = 0$ is clear. □

3.2 Cohen's structure theorems

Theorem 3.2.1

a) *Let C be a Cohen ring, B a complete noetherian local ring, and J an ideal of B. Then any local homomorphism $C \to B/J$ factors as a composite of local homomorphisms $C \to B \to B/J$.*

b) *Let K be a field. There exist a Cohen ring C with residue field isomorphic to K. Any two such Cohen rings are isomorphic by an isomorphism inducing the identity map in the residue fields.*

Proof a) (3.1.2) and (2.2.3). b) (3.1.8). □

Lemma 3.2.2 *Let $u \colon (C, \mathfrak{m}, K) \to (B, \mathfrak{n}, L)$ be a local homomorphism of complete noetherian local rings, which induces an isomorphism between the residue fields K and L. Let $\{x_1, \ldots, x_m\}$ be a set of generators of the maximal ideal $\mathfrak{m}$ of C, and $y_1, \ldots, y_n \in \mathfrak{n}$. Let $R = C[\![Y_1, \ldots, Y_n]\!]$. Then*

a) *There exists a unique ring homomorphism $v \colon R \to B$ extending u and such that $v(Y_i) = y_i$ for all i.*

b) *v is surjective if and only if $\{u(x_1), \ldots, u(x_m), y_1, \ldots, y_n\}$ generates the ideal $\mathfrak{n}$.*

c) *v is finite if and only if $\{u(x_1), \ldots, u(x_m), y_1, \ldots, y_n\}$ generates an ideal of definition of B.*

Proof a) is clear and b) follows from (3.1.4).

c) The ideal generated by $\{u(x_1), \ldots, u(x_m), y_1, \ldots, y_n\}$ is $v(\mathfrak{q})B$, where $\mathfrak{q}$ is the maximal ideal of R. We have that

$v(\mathfrak{q})B$ is an ideal of definition of B $\Leftrightarrow$ the length $L_B(B/v(\mathfrak{q})B) < \infty \Leftrightarrow^* L_{R/\mathfrak{q}}(B/v(\mathfrak{q})B) < \infty \Leftrightarrow \dim_{R/\mathfrak{q}}(B/v(\mathfrak{q})B) < \infty \Leftrightarrow^{**} v$ is finite,

where the equivalence marked with * is as follows:

Assume $L_B(B/v(\mathfrak{q})B) < \infty$. Let

$$B/v(\mathfrak{q})B = M_n \supset M_{n-1} \supset \cdots \supset M_0 \supset M_{-1} = 0$$

be a (maximal) composition series of the B-module $B/v(\mathfrak{q})B$. Then the B-modules M_i/M_{i-1} are simple, and so they are (simple) $B/\mathfrak{n}$-modules. Since $R/\mathfrak{q} = B/\mathfrak{n}$, the M_i/M_{i-1} are simple $R/\mathfrak{q}$-modules, and so $B/v(\mathfrak{q})B = M_n \supset M_{n-1} \supset \cdots \supset M_0 \supset M_{-1} = 0$ is a (maximal) composition series for the $R/\mathfrak{q}$-module $B/v(\mathfrak{q})B$. The converse is clear.

Finally, $\Rightarrow^{**}$ follows from (3.1.4) applied to a homomorphism $R^t \to B$ such that $R^t/\mathfrak{q}R^t \to B/v(\mathfrak{q})B$ is surjective, and $\Leftarrow^{**}$ is clear. $\qquad\square$

Lemma 3.2.3 *Let* $u\colon (C, \mathfrak{m}, K) \to (B, \mathfrak{n}, L)$ *be a local homomorphism of complete noetherian local rings, which induces an isomorphism between the residue fields K and L. Assume that C is regular and let* $\{x_1, \ldots, x_m\}$ *be a regular system of parameters of C. Let $y_1, \ldots, y_n \in \mathfrak{n}$, $R = C[\![Y_1, \ldots, Y_n]\!]$, and $v\colon R \to B$ extending u as in (3.2.2). Then*

a) $v\colon R \to B$ *is injective and finite if and only if $u(x_1), \ldots, u(x_m)$, $y_1, \ldots, y_n$ is a (not necessarily regular) system of parameters of B. In this case, $\dim B = m + n$.*

b) *If $u(x_1), \ldots, u(x_m), y_1, \ldots, y_n$ is a part of a (not necessarily regular) system of parameters of B, then v is injective.*

Proof a) The set $u(x_1), \ldots, u(x_m), y_1, \ldots, y_n$ is a system of parameters of B if and only if $v(\mathfrak{q}) = (u(x_1), \ldots, u(x_m), y_1, \ldots, y_n)$ is an ideal of definition of B and $\dim B = m + n$, and by (3.2.2) this is equivalent to v finite and $\dim B = m + n$. So we have to prove that if v is finite, then

$$\dim B = m + n \iff v \quad \text{is injective.}$$

If B is a finite R-algebra, $\dim B = \dim(R/\ker(v))$. If $\ker(v) \neq 0$, $\dim(R/\ker(v)) < \dim R = m + n$, since R is an integral domain. So if v is finite, $\dim B = m + n \iff v$ is injective.

b) If $u(x_1), \ldots, u(x_m), y_1, \ldots, y_n$ is a part of a system of parameters of B, let $y_{n+1}, \ldots, y_{n+r} \in \mathfrak{n}$ be such that $u(x_1), \ldots, u(x_m)$, $y_1, \ldots, y_n$, $y_{n+1}, \ldots, y_{n+r}$ is a system of parameters of B. There is a commutative diagram

$$R = C[\![Y_1, \ldots, Y_n]\!] \longrightarrow C[\![Y_1, \ldots, Y_n, Y_{n+1}, \ldots, Y_{n+r}]\!]$$

$$\searrow{\scriptstyle v} \qquad \swarrow{\scriptstyle w}$$

$$B$$

where the horizontal map is the inclusion, and w extends u by $w(Y_i) = y_i$ for $i = 1, \ldots, n + r$. By a), w is injective, and so v is injective.   □

Theorem 3.2.4 *Let $(B, \mathfrak{n}, K)$ be a complete noetherian local ring.*

a) *There exist a Cohen ring C and a surjective ring homomorphism $C[\![Y_1, \ldots, Y_n]\!] \to B$. If B contains a field, then there exists a surjective ring homomorphism $K[\![Y_1, \ldots, Y_n]\!] \to B$.*

b) *Suppose B is an integral domain or contains a field. Then there exists a finite injective local homomorphism $C[\![Y_1, \ldots, Y_m]\!] \to B$, where C is a Cohen ring (if B does not contain a field) or $C = K$ (if B contains a field).*

Proof a) Let C be a Cohen ring with residue field K (3.2.1), and $C \to B$ the homomorphism induced by $C \to K = B/\mathfrak{n}$ (3.2.1). By (3.2.2) there exists a surjective ring homomorphism $C[\![Y_1, \ldots, Y_n]\!] \to B$.

If B contains a field, it contains the field $\mathbb{F}_p$ ($\mathbb{F}_p = \mathbb{Z}/p\mathbb{Z}$ if $p > 0$, $\mathbb{F}_p = \mathbb{Q}$ if $p = 0$), where p is the characteristic of K. Therefore K is a formally smooth $\mathbb{F}_p$-algebra (2.4.6), and so by (2.2.3) there exists a section of the epimorphism $B \to K$. So B contains a field isomorphic to its residue field, and the proof continues as in the previous case.

b) If B contains a field, B contains a field isomorphic to K, as we have just seen. Then the result follows from (3.2.3). If B does not contain a field, then the characteristic of K is $p > 0$. Let C be a Cohen ring with residue field K, and $C \to B$ the homomorphism given by (3.2.1). The canonical homomorphism $\mathbb{Z}_{(p)} \to B$ is injective, since its kernel is a prime ideal of $\mathbb{Z}_{(p)}$ (B is a domain), which is not $(p)\mathbb{Z}_{(p)}$ (since B does not contain a field). So $p \cdot 1 \in B$ is a part of a system of parameters of B ([Mt, proof of Theorem 17.4.iii] or [ZS, VIII, §9, corollary 2 of Theorem 20]). By (3.2.3.b), $C \to B$ is injective. Now the result follows from (3.2.3.a).   □

Corollary 3.2.5 *Any complete noetherian local ring is a quotient of a complete regular local ring.*

Proof It follows from (3.2.4).   □

4

Complete intersections

The main purpose of this chapter is to prove the descent of the complete intersection property by flat local homomorphisms (4.3.8), which has as a consequence the localization theorem for complete intersections (4.3.9): if $(A, \mathfrak{m}, K)$ is a complete intersection ring, $\mathfrak{p}$ a prime ideal of A, then $A_{\mathfrak{p}}$ is complete intersection. This is another important result which appears without proof in Matsumura's book [Mt, end of Section 21]. The case when A is a quotient of a regular ring follows easily from the same localization property for regular rings (Serre's theorem). The difficult part, solved by Avramov [Av1], is to reduce the problem to this case.

We follow some papers by Avramov. We first need to present Gulliksen's proof [GL] of the existence of minimal DG algebra resolutions (4.1.7). This result is used to prove Main Lemma 4.2.1 following [Av1]. We do not know any easier proof of this lemma (or, equivalently, of (4.2.2)). Finally, we characterize complete intersection rings in terms of homology modules in order to prove the main theorems (4.3.8), (4.3.9). In (higher) André–Quillen homology theory, complete intersections are characterized by the vanishing of an H_3 module [An1, 6.27]. Since we want to avoid these higher homology modules, we characterize them by counting dimensions of the lower homology modules (4.3.5) as in [Av2, Section 3].

Avramov's Lemma 4.2.1 is very powerful (as an example we give in (4.4.2) an alternative proof of Kunz's characterization of regularity in characteristic p [Ku] using this lemma). For other applications and improvements of it see [Av2], [Av3].

4.1 Minimal DG resolutions

Definition 4.1.1 Let $A \to B$ be a surjective homomorphism of noetherian local rings. Let (R, d) be a free DG resolution of the A-algebra B with $R_0 = A$. Then (R, d) is said to be minimal if $d(R) \subset \mathfrak{m}R$, where $\mathfrak{m}$ is the maximal ideal of R_0.

Definition 4.1.2 Let (R, d) be a DG A-algebra. A derivation ϑ on R is an R_0-linear map $\vartheta \colon R \to R$ of homogeneous degree w and such that

    a) $d\vartheta = \vartheta d$
    b) $\vartheta(xy) = (-1)^{wq}\vartheta(x)y + x\vartheta(y)$ where $y \in R_q$.

Lemma 4.1.3 *Let ϑ be a derivation on a DG A-algebra R, and x a cycle in R. Let $R' = R \langle X; dX = x \rangle$. Then ϑ can be extended to a derivation ϑ' on R' if and only if $\vartheta(x) \in B(R')$.*

Proof If ϑ can be extended, $\vartheta(x) = \vartheta(d(X)) = d\vartheta'(X) \in B(R')$. Conversely, if $\vartheta(x) \in B(R')$, choose an element $G \in R'$ such that $d(G) = \vartheta(x)$.

If $\deg X$ is odd, we have

$$R' = R \oplus RX.$$

For $r, s \in R$, we define

$$\vartheta'(r + sX) = \vartheta(r) + (-1)^{\deg \vartheta}\vartheta(s)X + sG.$$

If $\deg X$ is even, we have

$$R' = \bigoplus_{i \geq 0} RX^{(i)}.$$

For $r_0, \ldots, r_m \in R$, we define

$$\vartheta'\left(\sum_{i=0}^{m} r_i X^{(i)}\right) = \sum_{i=0}^{m} \vartheta(r_i)X^{(i)} + \sum_{i=1}^{m} r_i X^{(i-1)}G.$$

The verification that ϑ' is well defined and becomes a derivation on R' is straightforward. $\square$

Definition 4.1.4 If ϑ', X and G are as above, we call ϑ' the *canonical extension* of ϑ satisfying $\vartheta'(X) = G$.

Example 4.1.5 Suppose that R' is a DG A-algebra of the form $R' = R\langle X; dX = x\rangle$. Then by (4.1.3) we may consider the canonical extension ϑ' of the trivial derivation (the zero map) on X satisfying $\vartheta'(X) = 1$. Explicitly,

- $\vartheta'(r + sX) = s$ if $\deg X$ is odd;
- $\vartheta'(\sum_{i=0}^{m} r_i X^{(i)}) = \sum_{i=1}^{m} r_i X^{(i-1)}$ if $\deg X$ is even.

This derivation ϑ' is called the derivation associated with the extension $R \to R'$.

Lemma 4.1.6 *Let $\vartheta\colon R \to R$ be a derivation on a DG A-algebra R. Let $\{x_i\}_{i \in I}$ be a set of cycles in R. Assume that there exist elements $G_i \in R$ such that $d(G_i) = \vartheta(x_i)$ for all $i \in I$. Let $R' = R\langle\{X_i\}_{i\in I}; dX_i = x_i\rangle$. Then ϑ has an extension to a derivation $\vartheta'\colon R' \to R'$ satisfying*

- *$\vartheta'(X_i) = G_i$ for all $i \in I$;*
- *$\vartheta'(X_i^{(m)}) = X_i^{(m-1)} G_i$ for all $i \in I$ and $m > 0$, if $\deg X_i$ is even.*

Such an extension ϑ' is unique.

Proof The uniqueness of ϑ' is clear. The existence of ϑ' follows from (4.1.3) in view of Definition 1.2.4 of $R\langle\{X_i\}_{i\in I}; dX_i = x_i\rangle$. $\qquad\square$

Theorem 4.1.7 *Let $(A, \mathfrak{m}, K)$ be a noetherian local ring. Then there exists a free DG resolution R of the A-algebra K with $R_0 = A$ which is minimal, i.e., such that $d(R) \subset \mathfrak{m}R$.*

Proof We use the construction of (1.2.6). We begin taking $R^0 = A$. At each step $n > 0$, we choose a minimal set of generators $\{c_{n,i}\}_{i\in I_n}$ of the R_0-module

$$\ker\big(H_{n-1}(R^{n-1}) \to H_{n-1}(K)\big)$$

$(= H_{n-1}(R^{n-1})$ if $n-1 > 0)$, and a set of homogeneous cycles $\{x_{n,i}\}_{i\in I_n}$ representing $\{c_{n,i}\}_{i\in I_n}$, and take $R^n = R^{n-1}\langle\{X_{n,i}\}_{i\in I_n}; dX_{n,i} = x_{n,i}\rangle$. Finally, we define $R = \varinjlim R^n$.

For each $n > 0$ and $i \in I_n$, we define $\vartheta^{n,i}$ as the unique derivation on R^n satisfying

$$\vartheta^{n,i}(R^{n-1}) = 0,$$

$$\vartheta^{n,i}(X_{n,i}) = 1 \quad \text{and} \quad \vartheta^{n,i}(X_{n,j}) = 0 \quad \text{if} \quad i \neq j,$$

and

$$\vartheta^{n,i}(X_{n,i}^{(m)}) = X_{n,i}^{(m-1)} \quad \text{and} \quad \vartheta^{n,i}(X_{n,j}^{(m)}) = 0 \quad \text{for } i \neq j$$

if $m > 0$ and $\deg X_{n,i}$ is even.

Such a derivation exists by (4.1.6). We will show that each of the derivations $\vartheta^{n,i}$ can be extended to a derivation $\theta^{n,i}$ on R. By a limit argument, it is enough to show the following: let $t \geq n$ and suppose that $\vartheta^{n,i}$ has been extended to a derivation ϑ on R^t. Then ϑ can be extended to a derivation ϑ' on R^{t+1}. We will prove this.

If $H_t(R^t) = 0$, then $R^{t+1} = R^t$, and so we put $\vartheta' = \vartheta$. Otherwise let $\{x_{t+1,j}\}_{j \in I_{t+1}}$ be the set of homogeneous cycles chosen above. The existence of ϑ' will follow from (4.1.6) if we show that

(∗) $\vartheta(x_{t+1,j}) \in B(R^t)$ for all $j \in I_{t+1}$.

Since $\deg \vartheta = -\deg X_{n,i} = -n < 0$, and $\deg x_{t+1,j} = t$, we have

$$\vartheta(x_{t+1,j}) \in Z_{t-n}(R^t).$$

Thus if $t > n$, (∗) is satisfied, since $H_q(R^t) = 0$ for $0 < q < t$. Hence we may assume that $t = n$. In this case the chosen cycles can be expressed as follows

(∗∗) $x_{t+1,j} = \sum_k r_{j,k} X_{n,k} + y_j,$

where $r_{j,k} \in R_0 = A, r_{j,k} \neq 0$ for only finitely many indices k, and $y_j \in R^{n-1}$. Differentiating (∗∗) yields $0 = \sum_k r_{j,k} x_{n,k} + dy_j$. Since $\{x_{n,i}\}_{i \in I_n}$ represents a minimal set of generators of $H_{n-1}(R^{n-1})$, it follows that $r_{j,k} \in \mathfrak{m}$ for all $k \in I_n$. From (∗∗) we obtain

$$\vartheta(x_{t+1,j}) = r_{j,i} \cdot 1 \in \mathfrak{m} R_0^n = B_0(R^t)$$

(the last equality follows since $t = n > 0$).

Hence for each $n > 0$ and $i \in I_n$, we have a derivation $\theta^{n,i} \colon R \to R$. We now show that

$$\bigcap_{n>0,\, i \in I_n} \ker(\theta^{n,i} \otimes_A K) = (R \otimes_A K)_0 = K.$$

For this, let $z \in \bigcap_{n>0,\, i \in I_n} \ker(\theta^{n,i} \otimes_A K)$. If $z \notin K$, we can choose $n > 0$ and $i \in I_n$, so that $z \in R^{n-1}\langle \{X_{n,j}\}_{j \in I_n}; dX_{n,j} = x_{n,j}\rangle$ and $z \notin R^{n-1}\langle \{X_{n,j}\}_{j \in I_n,\, j \neq i}; dX_{n,j} = x_{n,j}\rangle$.

We have the following commutative diagram with exact top row, where we abbreviate $dX_{n,j} = x_{n,j}$ by $dX = x$:

$$0 \to R^{n-1}\big\langle \{X_{n,j}\}_{\substack{j\in I_n \\ j\neq i}}; dX = x \big\rangle \to$$

$$R^{n-1}\big\langle \{X_{n,j}\}_{j\in I_n}; dX = x \big\rangle \to R^{n-1}\big\langle \{X_{n,j}\}_{j\in I_n}; dX = x \big\rangle$$

$$\downarrow \qquad\qquad\qquad\qquad \downarrow$$

$$R \qquad \xrightarrow{\ \theta^{n,i}\ } \qquad R$$

where the vertical maps are inclusions, and the right-hand map is the derivation associated with the extension

$$R^{n-1}\big\langle \{X_{n,j}\}_{\substack{j\in I_n \\ j\neq i}}; dX_{n,j} = x_{n,j} \big\rangle \to R^{n-1}\big\langle \{X_{n,j}\}_{j\in I_n}; dX_{n,j} = x_{n,j} \big\rangle,$$

(see (4.1.5)). Since the top row is a split exact sequence of free modules over R_0, we obtain a commutative diagram

$$0 \to R^{n-1}\big\langle \{X_{n,j}\}_{\substack{j\in I_n \\ j\neq i}}; dX = x \big\rangle \otimes_A K \to$$

$$R^{n-1}\big\langle \{X_{n,j}\}_{j\in I_n}; dX = x \big\rangle \otimes_A K \to R^{n-1}\big\langle \{X_{n,j}\}_{j\in I_n}; dX = x \big\rangle \otimes_A K$$

$$\downarrow \qquad\qquad\qquad\qquad\qquad \downarrow$$

$$R \otimes_A K \qquad \xrightarrow{\ \theta^{n,i} \otimes_A K\ } \qquad R \otimes_A K$$

where the top row is exact and similarly the vertical maps are injective. Since $z \in \ker(\theta^{n,i} \otimes_A K)$, we deduce that also

$$z \in R^{n-1}\big\langle \{X_{n,j}\}_{j\in I_n, j\neq i}; dX_{n,j} = x_{n,j} \big\rangle,$$

which is a contradiction.

Finally, we show by induction on q that $B_q(R \otimes_A K) = 0$. For $q = 0$ this is clear since $B_0(R) = \mathfrak{m}$. Let $p > 0$, and assume that $B_q(R \otimes_A K) = 0$ for $q < p$. For every $n > 0$ and $i \in I_n$, $\theta^{n,i} \otimes_A K$ is of negative degree and commutes with the differential on $R \otimes_A K$. Hence

$$(\theta^{n,i} \otimes_A K)(B_p(R \otimes_A K)) \subset \bigoplus_{q<p} B_q(R \otimes_A K) = 0.$$

Therefore $B_p(R \otimes_A K)$ is contained in

$$B_p(R \otimes_A K) \cap \Big(\bigcap_{\substack{n>0 \\ i\in I_n}} \ker(\theta^{n,i} \otimes_A K) \Big) = B_p(R \otimes_A K) \cap K = 0,$$

since $p > 0$. $\qquad\qquad\qquad\qquad\qquad\qquad\qquad\qquad\qquad\qquad\qquad \square$

4.2 The main lemma

We prove here the following key result of Avramov.

Lemma 4.2.1 *Let* $f\colon (A, \mathfrak{m}, K) \to (B, \mathfrak{n}, L)$ *be a flat local homomorphism of noetherian local rings. Let* $\{u_1, \ldots, u_n\}$ *be a minimal set of generators of the ideal* $\mathfrak{m}$ *of* A, *and* $\{u_1, \ldots, u_n, t_1, \ldots, t_m\}$ *a set of generators of the ideal* $\mathfrak{n}$ *of* B *(we denote an element of* A *and its image in* B *by the same symbol). Then the homomorphism induced by* f *between the first Koszul homology modules*

$$h\colon H_1(u_1, \ldots, u_n; A) \otimes_K L \to H_1(u_1, \ldots, u_n, t_1, \ldots, t_m; B)$$

is injective.

Proof Let Q be a minimal free DG resolution of the A-algebra K (4.1.7) having 1-skeleton $Q(1) = A\langle U_1, \ldots, U_n; dU_j = u_j \rangle$. Then $P := Q \otimes_A B$ is a free DG resolution of the B-algebra $K \otimes_A B$ (since f is flat) and minimal (since f is local).

Let g be the homomorphism

$$Z := P\langle T_1, \ldots, T_m; dT_i = 1 \otimes t_i \rangle \to K \otimes_A B\langle T_1, \ldots, T_m; dT_i = 1 \otimes t_i \rangle$$

obtained by applying $- \otimes_B B\langle T_1, \ldots, T_m; dT_i = 1 \otimes t_i \rangle$ to the map $P \to K \otimes_A B$. Since this last map induces isomorphisms in homology, so does g.

Let $L = B\langle U_1, \ldots, U_n, T_1, \ldots, T_m; dU_j = u_j, dT_i = t_i \rangle$ be the Koszul complex associated to the sequence $u_1, \ldots, u_n, t_1, \ldots, t_m$ in B. Let $M = L\langle V_1, \ldots, V_r; dV_i = v_i \rangle$ where $V_1, \ldots, V_r$ are the variables of degree 2 of the minimal resolution Q, and

$$v_i \in A\langle U_1, \ldots, U_n \rangle \subset B\langle U_1, \ldots, U_n, T_1, \ldots, T_m \rangle$$

their images by the differential of Q. Since the classes of the cycles v_i are a minimal set of generators of $H_1(Q(1)) = H_1(u_1, \ldots, u_n; A)$, to see that h is injective we have to see that the classes of $v_1, \ldots, v_r$ are linearly independent in $H_1(L) = H_1(u_1, \ldots, u_n, t_1, \ldots, t_m; B)$.

So suppose that the classes of $dV_1, \ldots, dV_r$ are linearly dependent in $H_1(L)$. This means that there is a $V = \sum r_i V_i$ with $r_i \in B$, some $r_j \notin \mathfrak{n}$, and an $S \in L_2$ such that $dV = dS$. Let $W_\alpha = U_\alpha$ for $1 \le \alpha \le n$, $W_\alpha = T_{\alpha-n}$ for $n < \alpha \le n + m$. Let $S = \sum_{\substack{\alpha, \beta=1 \\ \alpha < \beta}}^{n+m} b_{\alpha\beta} W_\alpha W_\beta$ with

$b_{\alpha\beta} \in B$. Define

$$(V - S)^{(k)} = \sum_{\gamma_k, \varepsilon_{\alpha,\beta}} \left(\prod_{i=1}^{r} r_i^{\gamma_i} V_i^{(\gamma_i)} \right) \left(\prod_{\substack{\alpha,\beta=1 \\ \alpha<\beta}}^{n+m} (b_{\alpha\beta} W_\alpha W_\beta)^{\varepsilon_{\alpha\beta}} \right),$$

where the sum runs over all $\gamma_k \geq 0$, $\varepsilon_{\alpha,\beta}$ such that

$$0 \leq \varepsilon_{\alpha\beta} \leq 1 \quad \text{and} \quad \sum_{i=1}^{r} \gamma_i + \sum_{\substack{\alpha,\beta=1 \\ \alpha<\beta}}^{n+m} \varepsilon_{\alpha\beta} = k.$$

Observe that this definition extends divided powers on the variables of even degree to the element of even degree $V - S$, using, as in the proof of (1.2.10), the rules

$$(x + y)^{(k)} = \sum_{i+j=k} x^{(i)} y^{(j)}, \quad \text{and}$$

$$(xy)^{(k)} = \begin{cases} x^k y^{(k)} & \text{if } \deg x \text{ is even, and } \deg y \text{ is even and } \geq 2, \\ 0 & \text{if } \deg x \text{ or } \deg y \text{ is odd and } k \geq 2. \end{cases}$$

We deduce from this that $d((V - S)^{(k)}) = (V - S)^{(k-1)} d(V - S) = 0$, and so $(V - S)^{(k)}$ is a cycle for all $k \geq 1$. The products of the elements T_i, U_j and $V_q^{(s)}$ for $s > 0$ form part of a basis of the free B-module Z, and in the decomposition of $(V - S)^{(k)}$ with respect to this basis, there is a summand $r_j^k V_j^{(k)} \notin \mathfrak{m} Z$. Since $dZ \subset \mathfrak{m} Z$, we deduce that the homology class of $(V - S)^{(k)}$ in Z is not zero for all $k > 0$. But as we have seen, $H_k(Z) = H_k(K \otimes_A B \langle T_1, \ldots, T_m; dT_i = 1 \otimes t_i \rangle)$, which is obviously zero for $k > m$. This is a contradiction. $\qquad\square$

Corollary 4.2.2 *Let* $(A, \mathfrak{m}, K) \rightarrow (B, \mathfrak{n}, L)$ *be a flat local homomorphism of noetherian local rings. Then the canonical homomorphism*

$$H_2(A, K, L) \rightarrow H_2(B, L, L)$$

is injective.

Proof This follows from (4.2.1) and the exact sequences (2.5.1):

$$\begin{array}{ccccc}
0 & \rightarrow & H_2(A, K, L) & \rightarrow & H_1(a_1, \ldots, a_m; A) \otimes_K L \rightarrow \cdots \\
 & & \downarrow & & \downarrow \\
0 & \rightarrow & H_2(B, L, L) & \rightarrow & H_1(b_1, \ldots, b_n; B) \qquad \rightarrow \cdots \quad \square
\end{array}$$

4.3 Complete intersections

Definition 4.3.1 We say that a noetherian local ring $(A, \mathfrak{m}, K)$ is a *complete intersection* if its $\mathfrak{m}$-adic completion can be given as a quotient of a regular local ring by an ideal generated by a regular sequence.

Definition 4.3.2 Let $(A, \mathfrak{m}, K)$ be a noetherian local ring. We define the complete intersection defect of A as the integer (1.4.4)

$$d(A) = \delta_2(A) - \delta_1(A) + \dim(A),$$

where $\delta_i(A) = \dim_K H_i(A, K, K)$ for $i = 1, 2$.

Lemma 4.3.3 *Let* $(A, \mathfrak{m}, K)$ *be a noetherian local ring,* $\widehat{A}$ *its ($\mathfrak{m}$-adic) completion. Then the canonical homomorphisms*

$$H_n(A, K, K) \to H_n(\widehat{A}, K, K)$$
$$H^n(\widehat{A}, K, K) \to H^n(A, K, K)$$

are isomorphisms for $n = 0, 1, 2$.

Proof $n = 0$ is clear since all modules are zero. Let $1 \le n \le 2$. For any positive integer s the ring homomorphisms

$$A \to A/\mathfrak{m}^s \to A/\mathfrak{m} = K$$

induce an exact sequence

$$H^{n-1}(A, A/\mathfrak{m}^s, K)$$
$$\to H^n(A/\mathfrak{m}^s, K, K) \to H^n(A, K, K) \to H^n(A, A/\mathfrak{m}^s, K).$$

Taking direct limits in $s > 0$, by (2.3.4) we obtain that the canonical homomorphism

$$\varinjlim H^n(A/\mathfrak{m}^s, K, K) \to H^n(A, K, K)$$

is an isomorphism. The same argument applied to the local noetherian ring $(\widehat{A}, \widehat{\mathfrak{m}}, K)$ shows that the canonical homomorphism

$$\varinjlim H^n(\widehat{A}/\widehat{\mathfrak{m}}^s, K, K) \to H^n(\widehat{A}, K, K)$$

is an isomorphism. The canonical isomorphisms $A/\mathfrak{m}^s = \widehat{A}/\widehat{\mathfrak{m}}^s$ show that the map $H^n(\widehat{A}, K, K) \to H^n(A, K, K)$ is an isomorphism.

The isomorphisms $H_n(A, K, K) \to H_n(\widehat{A}, K, K)$ thus follow from (1.4.5). $\qquad\square$

Proposition 4.3.4 *Let $(A, \mathfrak{m}, K)$ be a noetherian local ring. We have:*

a) $d(A) = d(\widehat{A})$.

b) *If $A = R/I$ where $(R, \mathfrak{n}, K)$ is a regular local ring, then*

$$d(A) = \dim_K(I/\mathfrak{n}I) - (\dim(R) - \dim(A)).$$

c) $d(A) \geq 0$.

Proof a) follows from (4.3.3). b) From the exact sequence

$$0 = H_2(R, K, K) \quad \rightarrow H_2(A, K, K)$$
$$\rightarrow H_1(R, A, K) = I/\mathfrak{n}I \rightarrow H_1(R, K, K) = \mathfrak{n}/\mathfrak{n}^2 \rightarrow H_1(A, K, K) \rightarrow 0$$

we obtain

$$\delta_2(A) - \delta_1(A) = \dim_K(I/\mathfrak{n}I) - \dim_K(\mathfrak{n}/\mathfrak{n}^2) = \dim_K(I/\mathfrak{n}I) - \dim(R).$$

c) By a), it is enough to show that $d(\widehat{A}) \geq 0$. By (3.2.5), $\widehat{A} = R/I$ where R is a regular local ring, and so by b), $d(\widehat{A}) = \dim_K(I/\mathfrak{n}I) - \operatorname{ht} I \geq 0$. $\qquad \square$

Proposition 4.3.5 *Let A be a noetherian local ring. Then A is complete intersection if and only if $d(A) = 0$.*

Proof It follows from (4.3.4) (see [Mt, Theorem 17.4]). $\qquad \square$

Proposition 4.3.6 *Let A be a noetherian local ring, and R any regular local ring such that $A = R/I$. Then A is complete intersection if and only if I is generated by a regular sequence.*

Proof (4.3.4), and (4.3.5). $\qquad \square$

Proposition 4.3.7 *Let $(A, \mathfrak{m}, K) \rightarrow (B, \mathfrak{n}, L)$ be a flat local homomorphism of noetherian local rings. Then*

$$d(B) = d(A) + d(B \otimes_A K).$$

Proof Bearing in mind that $H_n(A, K, L) = H_n(B, B \otimes_A K, L)$ by (1.4.3) and using (4.2.2), the Jacobi–Zariski exact sequence associated to $B \rightarrow B \otimes_A K \rightarrow L$

$$0 \rightarrow H_2(A, K, L) \rightarrow H_2(B, L, L) \rightarrow H_2(B \otimes_A K, L, L) \rightarrow$$
$$H_1(A, K, L) \rightarrow H_1(B, L, L) \rightarrow H_1(B \otimes_A K, L, L) \rightarrow 0$$

and the equality $\dim(B) = \dim(A) + \dim(B \otimes_A K)$, give the desired result. □

Theorem 4.3.8 *Let $(A, \mathfrak{m}, K) \to (B, \mathfrak{n}, L)$ be a flat local homomorphism of noetherian local rings. Then B is complete intersection if and only if A and $B \otimes_A K$ are complete intersection.*

Proof (4.3.4.c), (4.3.5), (4.3.7). □

Theorem 4.3.9 *Let $(A, \mathfrak{m}, K)$ be a noetherian local ring, $\mathfrak{p}$ a prime ideal of A. If A is complete intersection, then $A_\mathfrak{p}$ is complete intersection.*

Proof If A is a quotient of a regular local ring R, then the result follows from (4.3.6) and Serre's theorem: if R is regular then $R_\mathfrak{q}$ is regular for any prime ideal $\mathfrak{q}$ of R. So by (3.2.5), the result is valid for $\widehat{A}$.

Let $\mathfrak{p}$ be a prime ideal of A, and let $\mathfrak{q}$ be a prime ideal of $\widehat{A}$ such that $\mathfrak{q} \cap A = \mathfrak{p}$ (such an ideal exists, since $A \to \widehat{A}$ is faithfully flat). Then applying (4.3.8) to the flat local homomorphism $A_\mathfrak{p} \to \widehat{A}_\mathfrak{q}$, we deduce that $A_\mathfrak{p}$ is complete intersection. □

4.4 Appendix: Kunz's theorem on regular local rings in characteristic p

Lemma 4.4.1 *Let A be a ring that contains a field of characteristic $p > 0$ and B an A-algebra. Let $\phi_A \colon A \to A$, $\phi(a) = a^p$ be the Frobenius homomorphism, $^\phi A$ the ring A considered as A-module via ϕ_A, and similarly ϕ_B, $^\phi B$. The map induced by ϕ_A, ϕ_B*

$$H_1(A, B, M) \to H_1(^\phi A, ^\phi B, M)$$

is zero for any $^\phi B$-module M.

Proof As in the proof of (2.4.7) we may assume that the homomorphism $A \to B$ is surjective with kernel I. Since the Frobenius homomorphism induces the zero map $I/I^2 \to I/I^2$ the result follows from (1.4.1.c). □

Theorem 4.4.2 *Let $(A, \mathfrak{m}, K)$ be a noetherian local ring containing a field of characteristic $p > 0$. Let $\phi \colon A \to A$ be the Frobenius homomorphism, and $^\phi A$ as above. The following are equivalent:*

a) A is a regular local ring.

b) $^{\phi}A$ is a flat A-module.

Proof a) $\Longrightarrow$ b) Let $x_1, \ldots, x_n$ be a regular sequence generating the ideal $\mathfrak{m}$. By (2.6.1) we have an exact sequence

$$H_2(A, K, {}^{\phi}K) \to H_2({}^{\phi}A, K \otimes_A {}^{\phi}A, {}^{\phi}K) \to \operatorname{Tor}_1^A(K, {}^{\phi}A) \otimes_{\phi A} {}^{\phi}K \to$$
$$H_1(A, K, {}^{\phi}K) \to H_1({}^{\phi}A, K \otimes_A {}^{\phi}A, {}^{\phi}K) \to 0.$$

We have that $\ker({}^{\phi}A \to K \otimes_A {}^{\phi}A)$ is generated by the regular sequence $x_1^p, \ldots, x_n^p$ [Mt, Theorem 16.1], and so $H_2({}^{\phi}A, K \otimes_A {}^{\phi}A, {}^{\phi}K) = 0$ by (2.5.2). On the other hand, the homomorphism

$$H_1(A, K, {}^{\phi}K) \to H_1({}^{\phi}A, K \otimes_A {}^{\phi}A, {}^{\phi}K)$$

can be identified with the $^{\phi}K$-module homomorphism

$$f \colon \mathfrak{m}/\mathfrak{m}^2 \otimes_K {}^{\phi}K \to \mathfrak{m}^{[p]}/\mathfrak{m} \cdot \mathfrak{m}^{[p]},$$

defined by $f((x + \mathfrak{m}^2) \otimes a) = ax^p + \mathfrak{m} \cdot \mathfrak{m}^{[p]}$, where $\mathfrak{m}^{[p]} = \langle x^p : x \in \mathfrak{m} \rangle_A$.

Since A is regular, $\operatorname{gr}_{\mathfrak{m}}(A) = K[X_1, \ldots, X_n]$. Via this isomorphism, the map f composed with the canonical homomorphism

$$\mathfrak{m}^{[p]}/\mathfrak{m} \cdot \mathfrak{m}^{[p]} \to \mathfrak{m}^p/\mathfrak{m}^{p+1}$$

identifies to

$$g \colon K[X_1, \ldots, X_n]_1 \to K[X_1, \ldots, X_n]_p,$$

defined by $g(q(X_1, \ldots, X_n)) = q(X_1^p, \ldots, X_n^p)$, where $K[X_1, \ldots, X_n]_i$ is the ith homogeneous component of the polynomial ring $K[X_1, \ldots, X_n]$. This map g is clearly injective, and so f is injective.

Therefore from the exact sequence we obtain $\operatorname{Tor}_1^A(K, {}^{\phi}A) \otimes_{\phi A} {}^{\phi}K = 0$. Since $\operatorname{Tor}_1^A(K, {}^{\phi}A)$ is a $^{\phi}A$-module of finite type, Nakayama's lemma says that $\operatorname{Tor}_1^A(K, {}^{\phi}A) = 0$, and therefore $^{\phi}A$ is a flat A-module by the local flatness criterion [Mt, Theorem 22.3].

b) $\Longrightarrow$ a) Let $E = F_p$ be the prime subfield. By (1.4.6), (2.4.1), (2.4.5), we can identify $\alpha \colon H_2(A, K, {}^{\phi}K) \to H_2({}^{\phi}A, {}^{\phi}K, {}^{\phi}K)$ with the homomorphism $H_1(E, A, {}^{\phi}K) \to H_1({}^{\phi}E, {}^{\phi}A, {}^{\phi}K)$ which is zero by (4.4.1). But α is injective by (4.2.2). So $H_2(A, K, {}^{\phi}K) = 0$ and then A is regular by (2.5.3), (1.4.5). $\qquad\square$

Remarks

i) The maps $H_n(A, B, M) \to H_n({}^{\phi}A, {}^{\phi}B, M)$ as in (4.4.1) are zero also for $n = 0, 2$. For $n = 0$ it is clear from (1.4.1.b), since $db^p \otimes m = pb^{p-1}db \otimes m = 0$ in $\Omega_{\phi B | \phi A} \otimes_{\phi B} M$. For $n = 2$, note that by (1.4.1.d), (1.4.6) we can assume that $A \to B$ is surjective. If $I = \ker(A \to B)$ and $H_1(I)$ is the first Koszul homology module associated to a set of generators of I, we have an injection $H_2(A, B, M) \to H_1(I) \otimes_B M$ (same proof as (2.5.1)). So it is enough to show that the Frobenius homomorphism induces the zero map in $H_1(I)$. Let $\{a_i\}$ be a set of generators of I, $\alpha, \beta \colon A[\{X_i\}] \to A$ the homomorphisms of A-algebras defined by $\alpha(X_i) = 0$, $\beta(X_i) = a_i$. As in the proof of (2.5.4), $H(I) = \mathrm{Tor}_1^{A[\{X_i\}]}({}^{\alpha}A, {}^{\beta}A)$ (the fact that the number of variables in (2.5.4) is finite is not necessary). Applying $- \otimes_{A[\{X_i\}]} {}^{\beta}A$ to the exact sequence

$$0 \to (X_i) \to A[\{X_i\}] \to {}^{\alpha}A \to 0$$

we obtain an exact sequence

$$0 \to \mathrm{Tor}_1^{A[\{X_i\}]}({}^{\alpha}A, {}^{\beta}A) \to \frac{(X_i)}{(X_i)(X_i - a_i)} \to \frac{A[\{X_i\}]}{(X_i - a_i)}$$

and so $H_1(I) = \frac{(X_i) \cap (X_i - a_i)}{(X_i)(X_i - a_i)}$. If $z \in (X_i) \cap (X_i - a_i)$, then $z^p \in (X_i)(X_i - a_i)$ and so the result follows.

  ii) It is clear that the maps $H_n(A, B, M) \to H_n({}^{\phi}A, {}^{\phi}B, M)$ are zero for all $n \in \mathbb{N}$ by using simplicial resolutions [An3, Lemma 53].

iii) $a) \Rightarrow b)$ can be also easily deduced from [Mt, Theorem 23.1].

5

Regular homomorphisms:
Popescu's theorem

There are two important results in this chapter. The main one is that any regular homomorphism is a filtered inductive limit of finite type smooth homomorphisms. We will present here Popescu's proof ([Po1]-[Po3]), taking into consideration some simplifications by Ogoma, André and Swan ([Og2], [An4], [Sw]). In fact, our exposition will follow closely Swan's paper, which incorporates some ideas of all these papers.

In order to be consistent with the flavour of this book and also to show examples of the usefulness of these homological tools in Commutative Algebra, in some places where we can choose homological versus non homological methods we have preferred the former. This choice is also justified by the fact that nonhomological proofs are already available in literature (compare, e.g., the proof of Proposition 5.5.8 with [Bo, Chapter IX, §2.2], or that of Lemma 5.4.3 with [Mt, Theorem 23.1]).

The interested reader should notice that a different proof of Popescu's theorem was given by Spivakovsky in [Sp].

The second important result in this chapter is Theorem 5.6.1: the module of differentials of a regular homomorphism is flat. This result is due to André in 1974 [An1, Supplément, Théorème 30]. In homological terms it means that if $A \to B$ is a homomorphism of noetherian rings, the vanishing of $H_1(A, B, B_{\mathfrak{q}}/\mathfrak{q}B_{\mathfrak{q}})$ for all prime ideals $\mathfrak{q}$ of B implies the vanishing of $H_1(A, B, M)$ for any B-module M (see the proof of Proposition 5.6.2). Though this statement involves only H_1, André's proof uses H_n for all n in an essential way.

Alternatively, since 1986, we can deduce this result from Popescu's theorem. This is how we prove it in this book. However, the proof of Popescu's theorem is long and difficult. It would be desirable to find a simpler proof of (5.6.1).

5.1 The Jacobian ideal

Definition 5.1.1 Let A be a ring, M an A-module of finite presentation. Let $x = \{x_1, \ldots, x_n\}$ be a set of generators of M, F a free A-module with basis $\{e_1, \ldots, e_n\}$, $p \colon F \to M$ the homomorphism with $p(e_i) = x_i$. Let $y = \{y_1, \ldots, y_r\}$ be a set of generators of $N = \ker p$. Let

$$y_j = \sum_{i=1}^{n} a_{ij} e_i \quad a_{ij} \in A, \quad j = 1, \ldots, r.$$

Let s be an integer and let $\Delta_{x,s}$ be the ideal of A generated by the $(n-s) \times (n-s)$ minors of the matrix $\mathbf{a} = (a_{ij})$ (if $t \le 0$, we define the determinant of a $t \times t$ matrix as 1).

Lemma 5.1.2 *The ideal $\Delta_{x,s}$ does not depend on the choice of the set of generators y of N.*

Proof Let $y' = \{y_1', \ldots, y_t'\}$ be another set of generators of N,

$$y_j' = \sum_{i=1}^{n} a_{ij}' e_i \quad a_{ij}' \in A, \quad j = 1, \ldots, t.$$

For each $j = 1, \ldots, t$, let

$$y_j' = \sum_{k=1}^{r} b_{kj} y_k.$$

Then $\mathbf{a}\mathbf{b} = \mathbf{a}'$, where $\mathbf{b} = (b_{kj})$, etc. Therefore any $(n-s) \times (n-s)$ minor of $\mathbf{a}'$ is a linear combination of $(n-s) \times (n-s)$ minors of $\mathbf{a}$ (easy computation; see, e.g., Bourbaki, Algèbre, chapitre III, Ex.§8.6), and so $\Delta_{x,s}' \subset \Delta_{x,s}$. By symmetry, $\Delta_{x,s}' = \Delta_{x,s}$. $\qquad\square$

Lemma 5.1.3 *The ideal $\Delta_{x,s}$ depends only on M (and s) and not on the set of generators x of M.*

Proof It is enough to show that if $x' = \{x_1, \ldots, x_n, x_{n+1}\}$ with $x_{n+1} \in M$ a superfluous element, then $\Delta_{x,s} = \Delta_{x',s}$. Let $x_{n+1} = \sum_{i=1}^{n} b_i x_i$, $b_i \in A$. Let N, y, a_{ij}, be as in (5.1.1), and

$$0 \to N' \to F' \xrightarrow{p'} M \to 0$$

an exact sequence with F' a free A-module with basis $\{e_1, \ldots, e_n, e_{n+1}\}$, and $p'(e_i) = x_i$ for $i = 1, \ldots, n+1$. Consider the set of generators $y' = \{y_1', \ldots, y_r', -\sum_{i=1}^{n} b_i e_i + e_{n+1}\}$, where y_i' is the image of $y_i \in N$

in N' via the inclusion $F \to F'$ taking e_i to e_i. The matrix (a'_{ij}) of this presentation is

$$\begin{pmatrix} a_{11} & \cdots & a_{1r} & -b_1 \\ \vdots & & \vdots & \vdots \\ a_{n1} & \cdots & a_{nr} & -b_n \\ 0 & \cdots & 0 & 1 \end{pmatrix}.$$

The result follows easily using (5.1.2) and the fact that $\Delta_{x,s} \subset \Delta_{x,s+1}$ for any s. □

So we shall denote $\Delta_{M,s}$ instead of $\Delta_{x,s}$.

Example 5.1.4 If M is a free A-module of rank n, $\Delta_{M,n} = A$, $\Delta_{M,s} = 0$ for $s < n$.

Lemma 5.1.5 *If B is an A-algebra, then*

$$\Delta_{M,s}B = \Delta_{M \otimes_A B, s}.$$

Proof Clear. □

Lemma 5.1.6 *Let M_1, M_2 be A-modules and $M = M_1 \oplus M_2$. Then*

$$\Delta_{M,s} = \sum_{s_1+s_2=s} \Delta_{M_1,s_1} \Delta_{M_2,s_2}.$$

In particular, $\Delta_{M,s} = \Delta_{M \oplus A^m, s+m}$ for any m.

Proof Take a presentation direct sum of two presentations. The matrix is of the form

$$\begin{pmatrix} (a_{ij}) & 0 \\ 0 & (b_{kl}) \end{pmatrix}$$

and so the result follows easily. □

Lemma 5.1.7 *Let M, F, N, n, r, etc., be as in (5.1.1). Then $\Delta_{M,n-r} = A$ if and only if N is a free direct summand of F with basis $\{y_1, \ldots, y_r\}$.*

Proof Assume first that N is a free direct summand of F with basis $\{y_1, \ldots, y_r\}$. Then $M \oplus N = F$ is free. By (5.1.6) and (5.1.4) $\Delta_{M,n-r} = \Delta_{M \oplus N, n} = A$. Conversely, if $\Delta_{M,n-r} = A$, the set $\{y_1, \ldots, y_r\}$ is linearly independent, and so N is free with basis $\{y_1, \ldots, y_r\}$. By (5.1.5), for any A-algebra $B = A/I$, we have $\Delta_{M \otimes_A B, n-r} = B$, and so, by the same argument, the B-submodule generated by $\{y_1 \otimes 1, \ldots, y_r \otimes 1\}$ is free, hence the homomorphism $N \otimes_A B \to F \otimes_A B$ is injective. Therefore M is a flat A-module, and then, being of finite presentation, projective. Thus the exact sequence

$$0 \to N \to F \to M \to 0$$

is split.   □

Definition 5.1.8 Let A be a ring, $B = A[X_1, \ldots, X_n]/I$, where I is a finite type ideal of $A[X_1, \ldots, X_n]$, $C = S^{-1}B$, where S is a multiplicative subset of B. Let T be a multiplicative subset of $A[X_1, \ldots, X_n]$ such that $T^{-1}A[X_1, \ldots, X_n]/T^{-1}I = S^{-1}(A[X_1, \ldots, X_n]/I) = C$. For each finite type ideal $J \subset I$, let $B_J = A[X_1, \ldots, X_n]/J$, S_J the image of T in B_J.

Let $\Omega_{B_J|A}$ be the B_J-module of differentials, which is of finite presentation, and let $\Delta_{B_J|A} = \Delta_{\Omega_{B_J|A}, n-\mu(J)}$, where $\mu(J)$ is the minimum number of generators of J. We can compute $\Delta_{B_J|A}$ using the presentation

$$J/J^2 \to \Omega_{A[X_1,\ldots,X_n]|A} \otimes_{A[X_1,\ldots,X_n]} B_J \to \Omega_{B_J|A} \to 0.$$

That is, if $\{f_1, \ldots, f_{\mu(J)}\}$ is a minimal set of generators of J, $\Delta_{B_J|A}$ is the ideal of B_J generated by the $\mu(J) \times \mu(J)$ minors of the Jacobian matrix $(\partial f_i/\partial X_j)$.

Finally we define

$$H_{C|A}(B) = \mathrm{rad}(\sum_J \Delta_{B_J|A}(J : I)C),$$

where $(J : I) = \{f \in A[X_1, \ldots, X_n] : fI \subset J\}$.

If R is a multiplicative subset of C, then

$$R^{-1}H_{C|A}(B) = H_{R^{-1}C|A}(B).$$

Proposition 5.1.9 *Let $\mathfrak{q}$ be a prime ideal of C. The following are equivalent:*

    i) $C_\mathfrak{q}$ is a smooth A-algebra.
    ii) $H_{C|A}(B) \not\subset \mathfrak{q}$.

In particular, C is a smooth A-algebra if and only if $H_{C|A}(B) = C$.

Proof It is enough to show the last claim, and we can assume that C is local.

Suppose $H_{C|A}(B) = C$. Then, since C is local, there exists J such that $\Delta_{B_J|A}(J : I)C = C$. That means $\Delta_{B_J|A}C = C$ and $(J : I)C = C$.

Let $\mathfrak{m}$ be a maximal ideal of $T^{-1}A[X_1, \ldots, X_n]$ such that its image in C is the maximal ideal of this local ring. Then $(J : I)C = C$ implies

$$(T^{-1}J : T^{-1}I) = T^{-1}(J : I) \not\subset \mathfrak{m},$$

and so

$$((T^{-1}J)_\mathfrak{m} : (T^{-1}I)_\mathfrak{m}) = (T^{-1}J : T^{-1}I)_\mathfrak{m} = T^{-1}A[X_1, \ldots, X_n]_\mathfrak{m}$$

that is, $(T^{-1}J)_\mathfrak{m} = (T^{-1}I)_\mathfrak{m}$, and therefore, replacing T by the inverse image in $A[X_1, \ldots, X_n]$ of $T^{-1}A[X_1, \ldots, X_n] - \mathfrak{m}$, we can assume $T^{-1}J = T^{-1}I$. Thus $S_J^{-1}B_J = T^{-1}A[X_1, \ldots, X_n]/T^{-1}J = C$, where S_J is the image of T in B_J. Therefore, by the first equality $S_J^{-1}\Delta_{B_J|A} = \Delta_{B_J|A}S_J^{-1}B_J = \Delta_{B_J|A}C = C$ and (5.1.7), we have a split exact sequence

$$0 \to T^{-1}J/(T^{-1}J)^2 = S_J^{-1}(J/J^2)$$
$$\to \Omega_{A[X_1,\ldots,X_n]|A} \otimes_{A[X_1,\ldots,X_n]} C \to \Omega_{C|A} \to 0.$$

By (2.3.1), (1.4.7), C is a smooth A-algebra.

Conversely, assume that C is a smooth A-algebra. By (2.3.1) we have a split exact sequence

$$0 \to S^{-1}(I/I^2) \to \quad \Omega_{A[X_1,\ldots,X_n]|A} \otimes_{A[X_1,\ldots,X_n]} C \to \Omega_{C|A} \to 0,$$

and so, since C is local, $S^{-1}(I/I^2)$ is a free summand of the module $\Omega_{A[X_1,\ldots,X_n]|A} \otimes_{A[X_1,\ldots,X_n]} C$. Note that

$$\text{rank}\, S^{-1}(I/I^2) = \mu(T^{-1}I/(T^{-1}I)^2) = \mu(T^{-1}I)$$

by Nakayama's lemma. Let $r = \text{rank}\, S^{-1}(I/I^2)$ and let $y_1, \ldots, y_r \in I$ be elements such that their images in $T^{-1}I$ generate this ideal. Let $J = (y_1, \ldots, y_r)$, so that $T^{-1}J = T^{-1}I$ and so $(J : I)C = C$. The above exact sequence can be written as

$$0 \to T^{-1}J/(T^{-1}J)^2 \to \Omega_{A[X_1,\ldots,X_n]|A} \otimes_{A[X_1,\ldots,X_n]} C \to \Omega_{C|A} \to 0$$

and by (5.1.7), $\Delta_{\Omega_{B_J|A},n-\mu(J)}C = \Delta_{\Omega_{B_J|A},n-r}C = \Delta_{\Omega_{C|A},n-r} = C$. Therefore $\Delta_{B_J|A}(J : I)C = C$, and so $H_{C|A}(B) = C$. $\qquad\square$

Remark 5.1.10 From (5.1.9), we deduce that $H_{C|A}(B)$ does not depend on B, and so we will denote it simply by $H_{C|A}$.

Definition 5.1.11 Let C be an A-algebra essentially of finite presentation. Let $a \in H_{C|A}$. We say that a is *standard* with respect to a finite presentation $C = S^{-1}(A[X_1, \ldots, X_n]/I)$ if there exists a finite type ideal

$J \subset I$ such that $a \in \operatorname{rad}(\Delta_{B_J|A}(J:I)C)$. We say that a is standard if it is standard with respect to some finite presentation. We say that a is *strictly standard* if $a \in \Delta_{B_J|A}(J:I)C$.

Proposition 5.1.12 *In the notation as in (5.1.11), the following are equivalent:*

> i) a *is standard.*
> ii) a *is nilpotent, or* C_a *is a smooth A-algebra and* $\Omega_{C_a|A}$ *is a stably free* C_a-module.

Proof Assume that a is standard and not nilpotent. Let $J \subset I$ be an ideal of finite type such that $a \in \operatorname{rad}(\Delta_{B_J|A}(J:I)C)$. We have $\operatorname{rad}(\Delta_{B_J|A}(J:I)C)_a = C_a$, and so $\Delta_{B_J|A}(J:I)C_a = C_a$. As in the proof of (5.1.9), using (5.1.7) we deduce that $(S_J^{-1}J/J^2)_a$ is a free C_a-module and the exact sequence

$$0 \to (S_J^{-1}J/J^2)_a \to \Omega_{A[X_1,\ldots,X_n]|A} \otimes_{A[X_1,\ldots,X_n]} C_a \to \Omega_{C_a|A} \to 0$$

is split exact, and so $\Omega_{C_a|A}$ is a stably free C_a-module.

Assume now that C_a is a smooth A-algebra and $\Omega_{C_a|A}$ is stably free. Let $C_a = S^{-1}(A[X_1,\ldots,X_n]/I)$ for some S. We have a split short exact sequence (2.3.1)

$$0 \to S^{-1}(I/I^2) \to \Omega_{A[X_1,\ldots,X_n]|A} \otimes_{A[X_1,\ldots,X_n]} C_a \to \Omega_{C_a|A} \to 0.$$

Since $\Omega_{C_a|A}$ is a stably free C_a-module, so is $S^{-1}(I/I^2)$, that is, $S^{-1}(I/I^2) \oplus (C_a)^r$ is a free C_a-module. Replacing the presentation $A[X_1,\ldots,X_n]/I$ by $A[X_1,\ldots,X_n,Y_1,\ldots,Y_r]/(I+(Y_1,\ldots,Y_r))$, we obtain a similar exact sequence with the same term on the right and where the middle term is the direct sum of the former middle term and $(C_a)^r$. So this exact sequence has $S^{-1}(I/I^2) \oplus (C_a)^r$ as the left term. Thus we can assume that $S^{-1}(I/I^2)$ is a free C_a-module. Now the proof concludes in the same way as that of (5.1.9). □

Proposition 5.1.13 *Let* $f: B \to C$ *be a homomorphism of A-algebras essentially of finite type. Then* $(H_{B|A}C) \cap H_{C|B} \subset H_{C|A}$.

Proof Let $\mathfrak{q}$ be a prime ideal of C such that $(H_{B|A}C) \cap H_{C|B} \not\subset \mathfrak{q}$. Then $H_{B|A} \not\subset f^{-1}\mathfrak{q} =: \mathfrak{p}$, and $H_{C|B} \not\subset \mathfrak{q}$, that is, $B_{\mathfrak{p}}$ is a smooth A-algebra and $C_{\mathfrak{q}}$ is a smooth B-algebra. So $C_{\mathfrak{q}}$ is a smooth A-algebra and then $H_{C|A} \not\subset \mathfrak{q}$. □

Proposition 5.1.14 *Let B be an A-algebra essentially of finite presentation, C an A-algebra and $D = B \otimes_A C$. Then*

$$H_{B|A}D \subset H_{D|C}.$$

Proof It follows from (5.1.9), (2.3.1), (1.4.3), (1.4.7). $\qquad\square$

Lemma 5.1.15 *Let A be a ring, M an A-module. Let $S_A(M)$ be the symmetric A-algebra on M. The following are equivalent:*

   i) $S_A(M)$ is a smooth A-algebra.
   ii) M is a projective A-module.

Proof $S_A(M)$ is smooth if and only if $H_1(A, S_A(M), S_A(M)) = 0$ and $\Omega_{S_A(M)|A} \ (= M \otimes_A S_A(M))$ is a projective $S_A(M)$-module.

ii) $\Longrightarrow$ i) follows easily from the definition of smoothness. Alternatively, using (2.3.1), if M is projective, there exists a free A-module F and a factorization of the identity map $M \to F \to M$, and so $S_A(M) \to S_A(F) \to S_A(M)$. Therefore the identity map factors as

$$H_1(A, S_A(M), S_A(M)) \to H_1(A, S_A(F), S_A(M))$$
$$\to H_1(A, S_A(M), S_A(M))$$

with middle term zero by (1.4.1.d). So $H_1(A, S_A(M), S_A(M)) = 0$ and $\Omega_{S_A(M)|A} = M \otimes_A S_A(M)$ is a projective $S_A(M)$-module by base change.

i) $\Longrightarrow$ ii) If $S_A(M)$ is a smooth A-algebra, $\Omega_{S_A(M)|A} = M \otimes_A S_A(M)$ is a projective $S_A(M)$-module, and so by base change, $M = M \otimes_A S_A(M) \otimes_{S_A(M)} A$ is a projective A-module. $\qquad\square$

Example 5.1.16 Suppose that $C = R^{-1}(A[X_1, \ldots, X_n])/I$ is a smooth A-algebra essentially of finite presentation. Then we have a split exact sequence

$$0 \to I/I^2 \to \Omega_{A[X_1,\ldots,X_n]|A} \otimes_{A[X_1,\ldots,X_n]} C \to \Omega_{C|A} \to 0.$$

Let $D = S_C(I/I^2)$. By (5.1.15), D is a smooth C-algebra and so a smooth A-algebra. So we have a split exact sequence (1.4.6), (2.3.1)

$$0 \to \Omega_{C|A} \otimes_C D \to \Omega_{D|A} \to \Omega_{D|C} \to 0$$

and then $\Omega_{D|A} = (\Omega_{C|A} \otimes_C D) \oplus \Omega_{D|C} = (\Omega_{C|A} \otimes_C D) \oplus (I/I^2 \otimes_C D) = \Omega_{A[X_1,\ldots,X_n]|A} \otimes_{A[X_1,\ldots,X_n]} D$ is a free D-module.

So let C be an A-algebra essentially of finite presentation. The above construction shows that there exists a C-algebra of finite presentation

D (since I/I^2 is a C-module of finite presentation) such that $H_{C|A}D \subset H_{D|A}$ and $\Omega_{D_a|A}$ is a free D_a-module for all $a \in H_{C|A}$ (since C_a is a smooth A-algebra and $D_a = S_{C_a}(I_a/I_a^2)$). In particular D has a presentation such that the image in D of any element $a \in H_{C|A}$ is standard (5.1.12). Moreover, every A-algebra homomorphism $C \to E$ factors as $C \to D \to E$.

5.2 The main lemmas

Lemma 5.2.1 *Let $f\colon A \to B$ be a homomorphism of noetherian rings, $a \in A$ an element such that*

$$(0 : a) = (0 : a^2), (0 : f(a)) = (0 : f(a)^2).$$

Let $\overline{A} = A/(a^2)$, $\widetilde{A} = A/(a)$, $\overline{B} = B/a^2B$, $\widetilde{B} = B/aB$ (denoting a instead of $f(a)$ when clear). Let $\overline{A} \to \overline{C} \to \overline{B}$ be a factorization with $\overline{C}$ an $\overline{A}$-algebra of finite type. Then, there exists a factorization $A \to D \to B$ with D an A-algebra of finite type and a homomorphism $\overline{C} \to \widetilde{D} := D/aD$ of $\overline{A}$-algebras making the diagram

$$
\begin{array}{ccc}
\overline{C} & \to & \overline{B} \\
\downarrow & & \downarrow \\
\widetilde{D} & \to & \widetilde{B}
\end{array}
$$

commute, and such that $\pi^{-1}(H_{\overline{C}|\overline{A}}\overline{B}) \subset \mathrm{rad}(H_{D|A}B)$, where $\pi\colon B \to \overline{B}$ is the canonical map.

Proof Let $\overline{C} = \overline{A}[X_1, \ldots, X_r]/\overline{I}$. Let I be an ideal of $A[X_1, \ldots, X_r]$ such that $a^2 \in I$ and $\overline{I} = I/a^2A[X_1, \ldots, X_r]$, so $\overline{C} = A[X_1, \ldots, X_r]/I$.

By (5.1.8) we can choose a finite number of elements $h_{ij} \in I (i \in \Gamma, j \in \Lambda_i)$, $w_i \in A[X_1, \ldots, X_r]$ $(i \in \Gamma)$ such that, denoting $\overline{h}_{ij}, \overline{w}_i$ their images in $\overline{I}$ and $\overline{A}[X_1, \ldots, X_r]$, and denoting $J_i = (a^2, \{h_{ij}\}_{j \in \Lambda_i})$ for each i, $\overline{J}_i = (\{\overline{h}_{ij}\}_{j \in \Lambda_i})$ its image in $\overline{I}$, we have

$$H_{\overline{C}|\overline{A}} = \mathrm{rad}\left(\sum_{i \in \Gamma} \overline{p}_i \overline{C}\right),$$

where for each i, $\overline{w}_i\overline{I} \subset \overline{J}_i$, $p_i := m_i w_i$, $\overline{p}_i := \overline{m}_i \overline{w}_i$, where each m_i is a $|\Lambda_i| \times |\Lambda_i|$ minor of the matrix $(\partial h_{ij}/\partial X_t)_{j,t}$, and $\overline{m}_i$ its image in $\overline{A}[X_1, \ldots, X_r]$.

Take a finite set of generators of I of the form $\{h_{ij}\}_{i \in \Gamma', j \in \Lambda_i}$, where

$\Gamma' = \Gamma \cup \{*\}$. Let $v\colon A[X_1,\ldots,X_r] \to B$ be an A-algebra homomorphism making the diagram

$$
\begin{array}{ccc}
A[X_1,\ldots,X_r] & \to & B \\
\downarrow & & \downarrow \\
\overline{C} & \to & \overline{B}
\end{array}
$$

commute.

We have $v(h_{ij}) \in a^2 B$. Let $\zeta_{ij} \in aB$ be such that $v(h_{ij}) = a\zeta_{ij}$. Let $\underline{Z} = \{Z_{ij}\}_{i\in\Gamma', j\in\Lambda_i}$ be a set of variables and let $g_{ij} = h_{ij} - aZ_{ij} \in A[X_1,\ldots,X_r,\underline{Z}]$.

For each $k \in \Gamma$, $p_k I \subset J_k = (a^2, \{h_{kj}\}_{j\in\Lambda_k})$. So for each $i \in \Gamma'$, $j \in \Lambda_i$, let

$$p_k h_{ij} = \sum_{j_0\in\Lambda_k} H_{ijkj_0} h_{kj_0} + a^2 G_{ijk},$$

where if $i = k$ we choose $H_{ijkj_0} = p_k \delta_{j,j_0}$ and $G_{ijk} = 0$.

Let $F_{ijk} = p_k Z_{ij} - \sum_{j_0\in\Lambda_k} H_{ijkj_0} Z_{kj_0} - aG_{ijk}$. So $F_{ijk} = 0$ when $i = k$, and

$$aF_{ijk} = p_k(h_{ij} - g_{ij}) - \sum_{j_0\in\Lambda_k} H_{ijkj_0}(h_{kj_0} - g_{kj_0}) - a^2 G_{ijk}$$

$$= -p_k g_{ij} + \sum_{j_0\in\Lambda_k} H_{ijkj_0} g_{kj_0} \in (g),$$

where (g) is the ideal generated by $\{g_{ij}\}_{i\in\Gamma', j\in\Lambda_i}$.

Therefore $F_{ijk} \in ((g) : a)$. Also, it is clear that $F_{ijk} \in (\underline{Z}) + (a)$.

Let $(F) = (\{F_{ijk}\}_{i,j,k})$, $T = (g) + (F)$, $D = A[X_1,\ldots,X_r,\underline{Z}]/T$. The homomorphism $v\colon A[X_1,\ldots,X_r] \to B$ extends to $A[X_1,\ldots,X_r,\underline{Z}]$ via $Z_{ij} \mapsto \zeta_{ij}$ and that one induces a homomorphism $D \to B$, since (g) goes to 0 in B, $F \subset ((g) : a) \cap ((\underline{Z}) + (a))$, and this intersection also goes to 0 in B, since if $\varphi \in ((g) : a)$, we have that $a\varphi \in (g)$ and so $av(\varphi) = v(a\varphi) = 0$; if moreover $\varphi \in (\underline{Z}) + (a)$, then $v(\varphi) \in aB$ (we have $v(Z_{ij}) = \zeta_{ij} \in aB$) and so $v(\varphi) \in aB \cap (0 : a) = 0$ since $(0 : a) = (0 : a^2)$ in B by assumption.

We also have $I = (\{h_{ij}\}_{i\in\Gamma', j\in\Lambda_i}) \subset (g) + (a)$ in $A[X_1,\ldots,X_r,\underline{Z}]$, and so we have a homomorphism $\overline{C} = A[X_1,\ldots,X_r]/I \to \widetilde{D} := D/aD$. Clearly, this homomorphism makes the required diagram commutative. Moreover, since $D_a = A_a[X_1,\ldots,X_r,\underline{Z}]/T = A_a[X_1,\ldots,X_r]$, we have $a \in H_{D|A}$.

Choose an ordering in the sets Γ', Λ_i (for all $i \in \Gamma'$), and let h be the column vector $\{h_{ij}\}_{i\in\Gamma',j\in\Lambda_i}$ with the order $(i,j) \leq (i',j')$ if $(i < i')$ or $(i = i', j \leq j')$. Let $k \in \Gamma$. We can write the equality

$$p_k h_{ij} = \sum_{j_0\in\Lambda_k} H_{ijkj_0} h_{kj_0} + a^2 G_{ijk}$$

as

$$p_k h_{ij} = \sum_{\substack{i_0 \in \Gamma' \\ j_0 \in \Lambda_{i_0}}} H_{ijki_0j_0} h_{i_0j_0} + a^2 G_{ijk},$$

where

$$H_{ijki_0j_0} = \begin{cases} H_{ijkj_0} & \text{if } i_0 = k, \\ 0 & \text{if } i_0 \neq k, \end{cases}$$

and so, if H_k is the square matrix $(H_{ijki_0j_0})_{(i\in\Gamma',j\in\Lambda_i),(i_0\in\Gamma',j_0\in\Lambda_{i_0})}$ (here ij refers to the row and i_0j_0 to the column), we have

$$p_k h \equiv H_k h \quad \mathrm{mod}\ a.$$

Let

$$S_{k_1,k_2} = p_{k_2} H_{k_1} - H_{k_2} H_{k_1}.$$

We have

$$S_{k_1,k_2} h \equiv p_{k_2} p_{k_1} h - H_{k_2} p_{k_1} h \equiv p_{k_2} p_{k_1} h - p_{k_2} p_{k_1} h = 0 \quad \mathrm{mod}\ a.$$

So, since $a \in A$, differentiation in each coordinate gives

$$S_{k_1,k_2}(\partial h/\partial X_t) \in (a,h) = (a,g) \quad \text{for any } t = 1,\ldots r,$$

the meaning of this relation being that all coordinates of $S_{k_1,k_2}(\partial h/\partial X_t)$ are in the ideal of $A[X_1,\ldots,X_r,\underline{Z}]$ generated by the elements a, g_{ij}.

Since $p_{k_1} = m_{k_1} w_{k_1}$ with m_{k_1} a $|\Lambda_{k_1}| \times |\Lambda_{k_1}|$ minor of the matrix $(\partial h_{k_1j}/\partial X_t)_{j,t}$, reordering Γ' such that k_1 is the first element, and reordering $X_1,\ldots,X_r$ if necessary, we can suppose that $m_{k_1} = \det(M)$, where M is the $|\Lambda_{k_1}| \times |\Lambda_{k_1}|$-submatrix of the upper left corner of $(\partial h_{ij}/\partial X_t)_{(i,j),t}$.

Since $H_{ijk_1i_0j_0} = 0$ if $i_0 \neq k_1$, we have $(S_{k_1,k_2})_{(ij)(i_0j_0)} = 0$ for $i_0 \neq k_1$, and so we can write $(S_{k_1,k_2}) = (S|0)$, where S is the matrix formed by the first $|\Lambda_{k_1}|$ columns. Therefore we can write the relation

$$S_{k_1,k_2}(\partial h/\partial X_t) \in (a,g) \quad \text{for any } t = 1,\ldots r,$$

as

$$(S|0)\left(\frac{M}{*}\right) \in (a,g),$$

and so $SM \in (a, g)$.

Multiplying by the adjoint matrix of M, then by w_{k_1}, we have

$$p_{k_1} S_{k_1, k_2} \in (a, g).$$

Viewing $\underline{Z}$ as a column vector, $F_{(k_1)}$ the ideal generated by $\{F_{ijk_1}\}_{ij}$, and F_{k_2} the column vector $(F_{ijk_2})_{ij}$, we have $H_{k_1}\underline{Z} \equiv p_{k_1}\underline{Z} \mod (a, F_{(k_1)})$, and so

$$
\begin{aligned}
S_{k_1, k_2}\underline{Z} &= p_{k_2} H_{k_1}\underline{Z} - H_{k_2} H_{k_1}\underline{Z} \equiv p_{k_2} p_{k_1}\underline{Z} - p_{k_1} H_{k_2}\underline{Z} \\
&= p_{k_1}(p_{k_2}\underline{Z} - H_{k_2}\underline{Z}) \equiv p_{k_1} F_{k_2} \quad \mod (a, F_{(k_1)}).
\end{aligned}
$$

Let

$$E_{k_1} = \left(A[X_1, \ldots, X_r, \underline{Z}]/((g) + F_{(k_1)}) \right)_{p_{k_1}}.$$

We will show that E_{k_1} is a smooth A-algebra. Since

$$a F_{ijk_1} = -p_{k_1} g_{ij} + \sum_{j_0 \in \Lambda_{k_1}} H_{ijk_1 j_0} g_{k_1 j_0},$$

each g_{ij} is 0 in $\left(A[X_1, \ldots, X_r, \underline{Z}]/((\{g_{k_1 j}\}_{j \in \Lambda_{k_1}}) + F_{(k_1)}) \right)_{p_{k_1}}$, so that $E_{k_1} = \left(A[X_1, \ldots, X_r, \underline{Z}]/((\{g_{k_1 j}\}_{j \in \Lambda_{k_1}}) + F_{(k_1)}) \right)_{p_{k_1}}$. Also, $F_{ijk_1} = p_{k_1} Z_{ij} - \sum_{j_0 \in \Lambda_{k_1}} H_{ijk_1 j_0} Z_{k_1 j_0} - a G_{ijk_1}$, and so we can solve for the Z_{ij} and conclude that

$$E_{k_1} = \left(A[X_1, \ldots, X_r, \{Z_{k_1 j}\}_{j \in \Lambda_{k_1}}]/((\{g_{k_1 j}\}_{j \in \Lambda_{k_1}})) \right)_{p_{k_1}}.$$

Since $g_{ij} = h_{ij} - a Z_{ij}$, we have

$$\left(\partial g_{k_1 j}/\partial (X_t, Z_{k_1 j_0}) \right)_{j \in \Lambda_{k_1}, t=1, \ldots r, j_0 \in \Lambda_{k_1}} = \left((\partial h_{k_1 j}/\partial X_t)| - a\mathbf{I}_{|\Lambda_{k_1}|} \right).$$

By definition of p_{k_1}, some $|\Lambda_{k_1}| \times |\Lambda_{k_1}|$ minor of $(\partial h_{k_1 j}/\partial X_t)$ divides p_{k_1}, and so E_{k_1} is a smooth A-algebra.

In particular, E_{k_1} is a flat A-module, and so, applying $- \otimes_A E_{k_1}$ to the exact sequences $0 \to (0 : a) \to A \xrightarrow{a} A$ and $0 \to (0 : a^2) \to A \xrightarrow{a^2} A$, we see that $(0 : a) = (0 : a^2)$ in E_{k_1}. Since $p_{k_1} S_{k_1, k_2} \in (a, g)$, we see that $(S_{k_1, k_2})_{ij}$ goes to 0 in $E_{k_1}/a E_{k_1}$ and so F_{ijk_2} goes to 0 in $E_{k_1}/a E_{k_1}$, that is, $F_{ijk_2} \in a E_{k_1}$ for all i, j. We have $a F_{ijk_2} \in (g)$ and then $a F_{ijk_2} = 0$ in E_{k_1}. Since $(0 : a) = (0 : a^2)$ in E_{k_1}, $F_{ijk_2} = 0$ in E_{k_1}. Therefore $E_{k_1} = D_{p_{k_1}}$, showing that $p_{k_1} \in H_{D|A}$. Since $\pi^{-1}(H_{\overline{C}|\overline{A}}\overline{B})$ is contained in the radical of the ideal generated by a and by the p_k, we deduce $\pi^{-1}(H_{\overline{C}|\overline{A}}\overline{B}) \subset \mathrm{rad}(H_{D|A}B)$. $\qquad \square$

Lemma 5.2.2 *Let $f\colon A \to B$ be a homomorphism of noetherian rings, $a \in A$ such that $(0 : f(a)) = (0 : f(a)^2)$. Let $\widetilde{A} = A/(a^4)$, $\widetilde{B} = B/a^4 B$, etc. Let $A \to C \to B$ be a factorization of f with C an A-algebra of finite type. Assume that the image of a in C is strictly standard and that there exists a retraction of $\widetilde{A}$-algebras $\rho\colon \widetilde{C} \to \widetilde{A}$ making the diagram*

commute. Let $\mathfrak{a} = \left(0 : \frac{(0:a^2)_A}{(0:a)_A} \right)_A$. Then there exists a factorization $A \to C \to D \to B$ with D an A-algebra of finite type such that $\mathfrak{a} D \subset H_{D|A}$.

Proof Let $C = A[X_1, \ldots, X_r]/I$ be such that the image a' of a in C is strictly standard for this presentation, and for each i let $\widetilde{y}_i = \rho(\widetilde{X}_i) \in \widetilde{A}$, where $\widetilde{X}_i$ is the image of X_i in $\widetilde{C}$. Let y_i be a representant in A of $\widetilde{y}_i$. Let $J \subset I$ be an ideal such that $a' \in \Delta_{C_J|A}(J : I)C$, where $C_J = A[X_1, \ldots, X_r]/J$, and let $P(\underline{X}) \in \Delta_{C_J|A}(J : I)$ (where $\underline{X} = (X_1, \ldots, X_r)$, etc.) be a representant of a' in $A[\underline{X}]$, that is, $P(\underline{X}) - a' \in I$. Then $P(\widetilde{y}) - \widetilde{a} = 0$ in $\widetilde{A}$, and so $P(y) - a \in a^4 A$. In particular, $P(y) = as$ with $s \in A$ satisfying $s \equiv 1 \mod a$.

Let $P(\underline{X}) = \sum_v l_v m_v$ where the m_v are $u \times u$ minors of $(\partial \mathbf{f}/\partial \underline{X})$ where $u = \mu(J)$, and $\mathbf{f} = (f_1, \ldots, f_u)$ is a set of generators of J. Reordering the f_i and the X_t if necessary, suppose that

$$H_1 = \begin{pmatrix} \partial f_1/\partial X_1 & \cdots & & \partial f_1/\partial X_r \\ \vdots & & & \vdots \\ \partial f_u/\partial X_1 & \cdots & & \partial f_u/\partial X_r \\ 0 & & \cdots \ \mathbf{I}_{r-u} & \end{pmatrix}$$

is such that $\det H_1 = m_1$. For each v, we define H_v similarly, with the same first u rows, and changing the columns in the lower part of the matrix (in the $r - u$ lower rows) such that $\det H_v = m_v$.

Let G'_v be the adjoint matrix of H_v and $G_v(\underline{X}) = l_v G'_v$. Thus $G_v H_v = H_v G_v = l_v m_v \mathbf{I}_r$. We have then $P(\underline{X})\mathbf{I}_r = \sum_v G_v H_v$ and, in particular, $as\mathbf{I}_r = P(y)\mathbf{I}_r = \sum_v G_v(y) H_v(y)$.

Let ξ_i be the image of X_i in B and η_i the image of y_i in B. So $\xi_i \equiv \eta_i \mod a^4 B$, that is, $\xi_i = \eta_i + a^3 \varepsilon_i$ with $\varepsilon_i \in aB$. Let ε be the column vector $(\varepsilon_1, \ldots, \varepsilon_r)$ and similarly ξ and η. Let $\tau^v = (\tau^v_1, \ldots, \tau^v_r)$ be the column vector $H_v(\underline{y})\varepsilon$ ($\tau_i^v \in aB$ for all i). We have in B

$$\sum_v G_v(\underline{y})\tau^v = P(\underline{y})\varepsilon = as\varepsilon$$

and so

$$s(\xi - \eta) = sa^3\varepsilon = a^2 \sum_v G_v(\underline{y})\tau^v.$$

Since all the H_v have the same first u rows, the same is true for the τ^v. So we can write $\tau^v = (\tau_1, \dots, \tau_u, \tau_{u+1}^v, \dots, \tau_r^v)$.

Let $W_1, \dots, W_r$ be variables, and $T^v = (T_1, \dots, T_u, T_{u+1}^v, \dots, T_r^v)$ with T_j, T_j^v, variables. Let

$$h_j(\underline{X}, \underline{W}, \underline{T}) = s(X_j - y_j) - a^3 W_j - a^2(\sum_v G_v(\underline{y})T^v)_j \in A[\underline{X}, \underline{W}, \underline{T}].$$

Let

$$f_i(\underline{X}) - f_i(\underline{y}) = \sum_j \frac{\partial f_i}{\partial X_j}(\underline{y})(X_j - y_j) + \text{higher degree terms in } (X_j - y_j),$$

Since $s(X_j - y_j) \equiv a^3 W_j + a^2(\sum_v G_v(\underline{y})T^v)_j \mod h_j$, we see that if $m = \max\{\deg f_i\}$, then

$$s^m(f_i(\underline{X}) - f_i(\underline{y})) \equiv \sum_j s^{m-1}\frac{\partial f_i}{\partial X_j}(\underline{y})(a^3 W_j + a^2(\sum_v G_v(\underline{y})T^v)_j) + a^4 Q_i'$$

modulo (h), where $(h) = (\{h_j\})$, and

$$Q_i' \in (\underline{W}, \underline{T})^2 A[\underline{W}, \underline{T}].$$

So

$$s^m f_i(\underline{X}) - s^m f_i(\underline{y}) \equiv s^{m-1}a^2 \sum_j \frac{\partial f_i}{\partial X_j}(\underline{y})(\sum_v G_v(\underline{y})T^v)_j + a^3 Q_i \mod(h)$$

with $Q_i \in \sum_j A W_j + a(\underline{W}, \underline{T})^2 A[\underline{W}, \underline{T}]$.

Now $\sum_j \frac{\partial f_i}{\partial X_j}(\underline{y})(G_v(\underline{y})T^v)_j = (H_v(\underline{y})G_v(\underline{y})T^v)_i = (l_v(\underline{y})m_v(\underline{y})T^v)_i = l_v(\underline{y})m_v(\underline{y})T_i$ (since $i \leq u$), and so

$$\sum_j \frac{\partial f_i}{\partial X_j}(\underline{y})(\sum_v G_v(\underline{y})T^v)_j = \sum_v l_v(\underline{y})m_v(\underline{y})T_i = asT_i.$$

Therefore

$$s^m f_i(\underline{X}) - s^m f_i(\underline{y}) \equiv s^m a^3 T_i + a^3 Q_i \mod (h).$$

Since $f_i(\underline{y}) \in a^4 A$, let $f_i(\underline{y}) = a^3 c_i$ with $c_i \in aA$. Let $g_i = s^m c_i + s^m T_i + Q_i \in A[\underline{W}, \underline{T}]$. Then we have

$$
\begin{aligned}
a^3 g_i &= s^m a^3 c_i + s^m a^3 T_i + a^3 Q_i \\
&= s^m f_i(\underline{y}) + s^m a^3 T_i + a^3 Q_i \equiv s^m f_i(\underline{X}) \mod (h).
\end{aligned}
$$

Let $D = A[\underline{X}, \underline{W}, \underline{T}]/(I + (g, h)) = C[\underline{W}, \underline{T}]/(g, h)$, where (g) is the ideal generated by $\{g_i\}$ (in each of these rings). We have a canonical homomorphism of A-algebras $\varphi \colon A[\underline{X}, \underline{W}, \underline{T}] \to B$, sending X_j to ξ_j, W_j to 0, T_j to τ_j, T_j^v to τ_j^v. This homomorphism takes h and I to 0. So the above congruence mod (h) shows that $a^3 \varphi(g_i) = s^m \varphi(f_i(\underline{X})) = 0$. That is, $\varphi(g_i) \in (0 : a^3)_B = (0 : a)_B$. Since $\varphi(g_i) \in aB$ (by definition of g_i, since $c_i \in aA$, $\tau_j^v \in aB$, $Q_i \in \sum_j AW_j + a(\underline{W}, \underline{T})^2 A[\underline{W}, \underline{T}]$ and $\varphi(W_j) = 0$), we have $\varphi(g_i) = ab$ with $b \in B$, and $a\varphi(g_i) = 0$ implies $b \in (0 : a^2)_B = (0 : a)_B$. Therefore $\varphi(g_i) = ab = 0$, and so φ defines a homomorphism $D \to B$.

Next we see that $a \in H_{D|A}$. Since $a^3 g_i \equiv s^m f_i(\underline{X}) \mod (h)$, we have that g_i is 0 in $C_a[\underline{W}, \underline{T}]/(h)$, and so $D_a = C_a[\underline{W}, \underline{T}]/(h)$. We can solve in this ring the equations $h_i = 0$ for the W_i, and then $D_a = C_a[\underline{T}]$. So D_a is a smooth C_a-algebra, and therefore a smooth A-algebra, that is, $a \in H_{D|A}$.

By definition of h_j, $X_j \equiv y_j \mod (a^2, h_j)_s$ in $A_s[\underline{X}, \underline{W}, \underline{T}]$. In particular, if $q(\underline{X}) \in I$ then $q(\underline{X}) \equiv q(\underline{y}) \equiv 0 \mod (a^2, h)_s$ and then $I A_s[\underline{X}, \underline{W}, \underline{T}] \subset (a^2, h) A_s[\underline{X}, \underline{W}, \underline{T}]$. Similarly, $P(\underline{X}) \equiv P(\underline{y}) = as \mod (a^2, h)_s$, that is, for some λ, $s^\lambda P(\underline{X}) \equiv as^{\lambda+1} \mod (a^2, h)$ in $A[\underline{X}, \underline{W}, \underline{T}]$. Since $s \equiv 1 \mod a$, we have $as \equiv a \mod a^2$. Therefore $s^\lambda P(\underline{X}) \equiv a \mod (a^2, h)$ in $A[\underline{X}, \underline{W}, \underline{T}]$, and then $s^\lambda P(\underline{X}) \equiv ta \mod h$ in $A[\underline{X}, \underline{W}, \underline{T}]$ for some $t \in A[\underline{X}, \underline{W}, \underline{T}]$, $t \equiv 1 \mod a$ in $A[\underline{X}, \underline{W}, \underline{T}]$.

Let $E = A[\underline{X}, \underline{W}, \underline{T}]/(g, h)$. We have $D = E/IE$. From $a^3 g_i \equiv s^m f_i(\underline{X}) \mod h$, we deduce that $f_i = 0$ in E_s. Since $P(\underline{X})I \subset J = (f_1, \ldots, f_u)$, we have $atIE_s = 0$, and so $E_{ast} = D_{ast}$. The A-algebra D_a is smooth, and then so is $D_{ast} = E_{ast}$, that is, $ast \in H_{E|A}$. We shall show that, in fact, $st \in H_{E|A}$. In $E_s = A_s[\underline{X}, \underline{W}, \underline{T}]/(g, h)$ we can solve the equations $h_j = 0$ for the X_j so that $E_s = A_s[\underline{W}, \underline{T}]/(g)$. For each $i, j = 1, \ldots, u$, we have $\partial g_i/\partial T_j \equiv \delta_{ij} s^m \mod a$, since $Q_i \in \sum_j AW_j + a(\underline{W}, \underline{T})^2 A[\underline{W}, \underline{T}]$. So, taking determinants, we see that $s^{mu} + ae \in H_{E_s|A_s}$ for some $e \in E_s$. Thus $s^{mu+1} t + stae \in H_{E|A}$. Since $ast \in H_{E|A}$, we obtain $s^{mu+1} t \in H_{E|A}$, and then $st \in H_{E|A}$.

Since $I A_s[\underline{X}, \underline{W}, \underline{T}] \subset (a^2, h) A_s[\underline{X}, \underline{W}, \underline{T}]$, we have $IE_{st} \subset aE_{st}$ and then $IE_{st} = a\mathfrak{b}$, where $\mathfrak{b}$ is an ideal of E_{st}. Since $atIE_s = 0$, we have

$aIE_{st} = 0$ and so $a^2\mathfrak{b} = 0$. Hence

$$\mathfrak{b} \subset (0 : a^2)_{E_{st}} \subset (0 : a(0 : \frac{(0 : a^2)}{(0 : a)}))_{E_{st}} = (0 : a\mathfrak{a})_{E_{st}}$$

(denoting by a_0 the image of $a \in A$ in E_{st}, we have $\mathfrak{b} \subset (0 : a_0^2) \subset$
$(0 : a_0(0 : \frac{(0 : a_0^2)}{(0 : a_0)})) = (0 : a(0 : \frac{(0 : a^2)}{(0 : a)}))_{E_{st}} = (0 : a\mathfrak{a})_{E_{st}}$, where in
the middle equalities $\frac{(0 : a_0^2)}{(0 : a_0)} = \frac{(0 : a^2)_{E_{st}}}{(0 : a)_{E_{st}}}, \ldots$, we have used that E_{st},
being smooth over A, is a flat A-algebra).

Therefore we deduce $a\mathfrak{a}\mathfrak{b} = 0$, that is, $\mathfrak{a}IE_{st} = 0$, and so for any $\alpha \in \mathfrak{a}$,
$IE_{st\alpha} = 0$, and then $D_{st\alpha} = E_{st\alpha}$. Thus $st\alpha \in H_{D|A}$. Since $a \in H_{D|A}$
and s, t are $\equiv 1 \bmod a$, we see that $1 - st = \gamma a \in H_{D|A}$ for some γ, so
$\alpha - st\alpha \in H_{D|A}$, and then finally $\alpha \in H_{D|A}$. $\qquad\square$

Lemma 5.2.3 *Let A be a ring, $B = A[X_1, \ldots, X_n]/I$, where I is a finite
type ideal of $A[X_1, \ldots, X_n]$, $C = S^{-1}B$, where S is a multiplicative subset of B. Let A' be an A-algebra, $B' = B \otimes_A A'$, $I' = IA'[X_1, \ldots, X_n]$,
$C' = C \otimes_A A'$. If $a \in C$ is a strictly standard element for the presentation $C = S^{-1}(A[X_1, \ldots, X_n]/I)$, then the image of a in C' is strictly
standard for the presentation $C' = S^{-1}(A'[X_1, \ldots, X_n]/I')$.*

Proof Let $J \subset I$ be a finite type ideal such that $a \in \Delta_{B_J|A}(J : I)C$.
Let $J' = JA'[X_1, \ldots, X_n]$. By (5.1.5) we have

$$\Delta_{B_J|A}B'_{J'} = \Delta_{\Omega_{B_J|A}, n-\mu(J)}B'_{J'} = \Delta_{\Omega_{B'_{J'}|A'}, n-\mu(J)}$$

$$\subset \Delta_{\Omega_{B'_{J'}|A'}, n-\mu(J')} = \Delta_{B'_{J'}|A'}$$

since $\mu(J') \leq \mu(J)$. On the other hand, $(J : I)A'[X_1, \ldots, X_n] \subset (J' :
I')$, so $\Delta_{B_J|A}(J : I)C' \subset \Delta_{B'_{J'}|A'}(J' : I')C'$. $\qquad\square$

Lemma 5.2.4 *Let $f\colon A \to B$ be a homomorphism of noetherian rings,
$a \in A$ such that $(0 : a) = (0 : a^2)$, $(0 : f(a)) = (0 : f(a)^2)$. Let
$\widetilde{A} = A/(a^4)$, $\widetilde{B} = B/a^4B$, etc. Let $A \to C \to B$, $A \to D \to B$ be
factorizations of f where C and D are A-algebras of finite type. Assume
that there exists a homomorphism of $\widetilde{A}$-algebras $\varphi\colon \widetilde{C} \to \widetilde{D}$ making commutative the diagram*

*If the image of a in C is strictly standard, then there exists a factoriza-
tion $A \to E \to B$ of f with E an A-algebra of finite type, and A-algebra
homomorphisms $C \to E$, $D \to E$, such that the diagram*

$$
\begin{array}{ccccc}
C & \to & E & \leftarrow & D \\
 & \searrow & \downarrow & \swarrow & \\
 & & B & &
\end{array}
$$

is commutative and $H_{D|A}E \subset H_{E|A}$.

Proof Let $F = C \otimes_A D$. Let $\rho \colon \tilde{F} \to \tilde{D}$, $\rho(\tilde{c} \otimes \tilde{d}) = \varphi(\tilde{c})\tilde{d}$. Clearly,
composition with the canonical homomorphism $\tilde{D} \to \tilde{F}$ is the identity
of $\tilde{D}$, and the diagram

$$
\begin{array}{ccc}
\tilde{F} & \to & \tilde{D} \\
 & \searrow \quad \swarrow & \\
 & \tilde{B} &
\end{array}
$$

is commutative. Applying (5.2.2) to $D \to F \to B$ and to the image of
a in F (which is strictly standard in F by (5.2.3)), we see that there
exists $D \to F \to E \to B$ with $\mathfrak{a}E \subset H_{E|D}$, where $\mathfrak{a} = (0 : \dfrac{(0 : a^2)_D}{(0 : a)_D})_D$
(denoting also by a the image of a in D).

Let $x \in H_{D|A}$. Then D_x is smooth over A and in particular A-flat.
Therefore $(0 : a) = (0 : a^2)$ implies $(0 : a)_{D_x} = (0 : a^2)_{D_x}$. So $\mathfrak{a}_x = D_x$,
and then $E_x \subset (H_{E|D})_x = H_{E_x|D_x}$. Thus E_x is a smooth D_x-algebra
and so a smooth A-algebra, that is $x \in H_{E|A}$.   □

Lemma 5.2.5 *Let $A \to C \to B$ be homomorphisms of noetherian rings
with C an A-algebra of finite type. Let $a \in A$ be such that $(0 : a) = (0 :
a^2)$ in A and in B, and such that the image of a in C is strictly standard.
Let $c \geq 8$ be an integer and $\overline{A} = A/a^c A$, $\overline{B} = B/a^c B$, etc. Assume that
we have a factorization $\overline{A} \to \overline{C} \to \overline{F} \to \overline{B}$ with $\overline{F}$ an A-algebra of finite
type. Then, there exists a factorization $A \to C \to E \to B$ with E an
A-algebra of finite type such that $\pi^{-1}(H_{\overline{F}|\overline{A}}\overline{B}) \subset \mathrm{rad}(H_{E|A}B)$, where
$\pi \colon B \to \overline{B}$ is the canonical map.*

Proof Let $c = 8$. Let $\tilde{A} = A/a^4 A$, $\tilde{B} = B/a^4 B$, etc. By (5.2.1) (with
a^4 instead of a), there exists a factorization $A \to D \to B$ with D an A-
algebra of finite type and a homomorphism $\overline{F} \to \tilde{D}$, making a diagram
as in (5.2.1) commute, such that $\pi^{-1}(H_{\overline{F}|\overline{A}}\overline{B}) \subset \mathrm{rad}(H_{D|A}B)$.

The homomorphism $\overline{C} \to \overline{F}$ induces a homomorphism $\tilde{C} \to \tilde{F}$, and the homomorphism $\overline{F} \to \tilde{D}$ induces $\tilde{F} \to \tilde{D}$. So we have a homomorphism $\tilde{C} \to \tilde{D}$. Applying (5.2.4) to the factorizations $A \to C \to B, A \to D \to B$, we obtain a factorization $A \to C \to E \to B$ with E an A-algebra of finite type such that $H_{D|A}E \subset H_{E|A}$.

This proves the result for $c = 8$. If $c \geq 8$, by (5.1.14), $H_{\overline{F}|\overline{A}}\breve{F} \subset H_{\breve{F}|\breve{A}}$, where $\breve{A} = A/a^8 A, \breve{F} = F/a^8 F$, and so the result follows from the case $c = 8$. $\square$

5.3 Statement of the theorem

Definition 5.3.1 A homomorphism $A \to B$ of noetherian rings is *regular* if it is flat and the fibers $k(\mathfrak{p}) \to B \otimes_A k(\mathfrak{p})$ are geometrically regular for all prime ideals $\mathfrak{p}$ of A.

The main result we want to prove in this chapter is:

Theorem 5.3.2 *Let $f \colon A \to B$ be a regular homomorphism of noetherian rings. Then f is a filtered inductive limit of smooth homomorphisms of finite type.*

Theorem 5.3.3 *Let $f \colon A \to B$ be a regular homomorphism of noetherian rings. Assume that we have a factorization of f, $A \to C \xrightarrow{g} B$, where C is an A-algebra of finite type. Then there exists a factorization $C \to D \to B$ of g such that D is a smooth A-algebra of finite type.*

These two theorems are equivalent. For, if $f = \varinjlim_i (f_i \colon A \to D_i)$ with each f_i smooth of finite type, and $A \to C \xrightarrow{g} B$ is a factorization of f with $C = A[X_1, \ldots, X_n]/(h_1, \ldots, h_r)$, then the homomorphism $A[X_1, \ldots, X_n] \to B$ factors through some D_i, and we can take i large enough so that the images of the elements $h_1, \ldots, h_r$ are 0. So g factors through some D_i.

Conversely, let $f = \varinjlim_i (\phi_i \colon A \to C_i)$ with each ϕ_i of finite type. For each i, we factor $C_i \to B$ as $C_i \to D_i \to B$ with D a smooth A-algebra of finite type. So we have a homomorphism $\varinjlim_i D_i \to B$ which is surjective.

To see that it is also injective, since D_i is an A-algebra of finite type, reasoning as before, there exists some $j \geq i$ such that $D_i \to B$ factors as $D_i \to C_j \to B$. So if $x \in \varinjlim_i D_i$ goes to 0 in B, taking some i such that $x \in D_i$ and some $k \geq j$ such that x goes to 0 in C_k, we see that x is 0 in D_k, and so in $\varinjlim_i D_i$.

Lemma 5.3.4 *Let B be an A-algebra of finite presentation, S a multiplicative subset of B. If $S^{-1}B$ is a smooth A-algebra, then there exists some $u \in S$ such that B_u is a smooth A-algebra.*

Proof As we saw in (5.1.8), $S^{-1}H_{B|A} = H_{S^{-1}B|A} = S^{-1}B$. Therefore there exists $x \in H_{B|A}$, $s \in S$, such that $x/s = 1$ in $S^{-1}B$, that is, there exist $s, t \in S$ such that $(x - s)t = 1$ in B. Thus $(H_{B|A})_{st} = B_{st}$, i.e., $H_{B_{st}|A} = B_{st}$. $\qquad\qquad\square$

Lemma 5.3.5 *Let $A \to C \to B$ be ring homomorphisms with C an A-algebra of finite presentation. If there exists a factorization $A \to C \to E \to B$ with E a smooth A-algebra essentially of finite presentation, then there exists a factorization $A \to C \to D \to B$ with D a smooth A-algebra of finite presentation.*

Proof Let $E = S^{-1}F$, with F of finite presentation over A. By (5.3.4), there exists $s \in S$ such that F_s is a smooth A-algebra. Let $C = A[X_1, \ldots, X_n]/(f_1, \ldots, f_r)$. Since n and r are finite, there exists $t \in S$ such that $C \to E = S^{-1}F$ factors through $C \to F_t \to E = S^{-1}F$, and so $C \to E$ factors as $C \to D := F_{st} \to E$. $\qquad\qquad\square$

Lemma 5.3.6 *Let $A \to C \to B$ be homomorphisms of noetherian rings, with C an A-algebra of finite type.*

> i) *Let $a \in A$ be such that its image in B belongs to $\mathrm{rad}(H_{C|A}B)$. Then there exists a factorization $A \to C \to D \to B$ with D an A-algebra of finite type such that $H_{C|A}D \subset H_{D|A}$ and the image of a in D is standard with respect to a finite presentation of D. So there exists an integer s such that the image of a^s in D is strictly standard.*
>
> ii) *If $H_{C|A}B = B$, then there exists a factorization $C \to D \to B$ with D a smooth A-algebra of finite type.*

Proof i) Let r be an integer such that the image of a^r in B belongs to $H_{C|A}B$. If $H_{C|A} = (c_1, \ldots, c_n)$, write $a^r = \sum_i b_i c_i$ with $b_i \in B$. Let $E = C[X_1, \ldots, X_n]/(a^r - \sum_i c_i X_i)$, and let $\varphi \colon E \to B$ be the homomorphism of C-algebras that sends X_i to b_i. Since $E_{c_i} =$

$C_{c_i}[X_1, \ldots, X_{i-1}, X_{i+1}, \ldots, X_n]$, which is smooth over A, we see that each $c_i \in H_{E|A}$ and so $H_{C|A}E \subset H_{E|A}$. By definition of E, the image of a^r in E belongs to $H_{C|A}E$, and so to $H_{E|A}$, and since this ideal is radical, the image of a in E belongs to $H_{E|A}$.

By (5.1.16) applied to $A \to E$, we see that there exists an E-algebra of finite type D with $H_{E|A}D \subset H_{D|A}$, and the image of a in D is standard for some presentation of D. Moreover, the homomorphism $E \to B$ factors through D.

ii) Take $a = 1$ in i). $\qquad\qquad\qquad\qquad\qquad\qquad\qquad\square$

Theorem 5.3.7 *Let $f\colon A \to B$ be a homomorphism of noetherian rings, $A \to C \to B$ a factorization of f with C an A-algebra of finite type. Let $\mathfrak{q}$ be a prime ideal of B minimal among those prime ideals containing $H_{C|A}B$, and assume that $\mathfrak{p} = f^{-1}(\mathfrak{q})$ is also minimal over $f^{-1}(H_{C|A}B)$. Assume moreover that $A_\mathfrak{p} \to B_\mathfrak{q}$ is formally smooth. Then there exists $A \to C \to D \to B$ with D an A-algebra of finite type such that $H_{C|A}B \subset \mathrm{rad}(H_{D|A}B) \not\subset \mathfrak{q}$.*

We now see that Theorem 5.3.7 implies Theorem 5.3.3. Assume that (5.3.7) holds. Let $f\colon A \to B$ be a regular homomorphism of noetherian rings, $A \to C' \to B$ a factorization with C' an A-algebra of finite type. Since B is noetherian, we can choose a factorization $A \to C' \to C \to B$ with $\mathrm{rad}(H_{C|A}B)$ maximal. By (5.3.6.ii), in order to prove (5.3.3), we have to show that $H_{C|A}B = B$. If not, let $\mathfrak{q}$ be a prime ideal of B minimal among those containing $H_{C|A}B$. It is enough to get a factorization $A \to C \to D \to B$ with D an A-algebra of finite type such that $H_{C|A}B \subset \mathrm{rad}(H_{D|A}B) \not\subset \mathfrak{q}$. If $\mathfrak{a} := f^{-1}(H_{C|A}B)$ and $\mathfrak{p}$ is a minimal prime ideal over $\mathfrak{a}$, then we can choose $\mathfrak{q}$ minimal over $H_{C|A}B$ such that $f^{-1}(\mathfrak{q}) = \mathfrak{p}$, since if $\mathrm{rad}(H_{C|A}B) = \mathfrak{q}_1 \cap \cdots \cap \mathfrak{q}_n$, then $\mathrm{rad}(\mathfrak{a}) = f^{-1}(\mathfrak{q}_1) \cap \cdots \cap f^{-1}(\mathfrak{q}_n) \subset \mathfrak{p}$ and therefore some $f^{-1}(\mathfrak{q}_i) \subset \mathfrak{p}$. Thus (5.3.7) implies (5.3.3).

We prove Theorem 5.3.7 at the end of Sections 5.4 and 5.5.

Definition 5.3.8 Let $A \to C \to B$ be ring homomorphisms with C an A-algebra of finite type. Let $\mathfrak{q}$ be a prime ideal of B. We say that $A \to C \to B \supset \mathfrak{q}$ is *resolvable* if the conclusion of (5.3.7) holds, that is, if there exists a factorization $A \to C \to D \to B$ with D an A-algebra of finite type such that $H_{C|A}B \subset \mathrm{rad}(H_{D|A}B) \not\subset \mathfrak{q}$.

Proposition 5.3.9 *Let $A \to C \to B$ be homomorphisms of noetherian* rings, with C an A-algebra of finite type. Let $\mathfrak{q}$ be a prime ideal of B, and $a \in A$ such that its image in C is strictly standard. If $H_{C|A}B \not\subset \mathfrak{q}$, then $A \to C \to B \supset \mathfrak{q}$ is (trivially) resolvable.

If $H_{C|A}B \subset \mathfrak{q}$ (and in particular the image of a in B belongs to $\mathfrak{q}$), let $c \geq 8$ be an integer and $\overline{A} = A/a^cA$, $\overline{B} = B/a^cB$, $\overline{\mathfrak{q}} = \mathfrak{q}/a^cB$, etc. Assume that $(0 : a) = (0 : a^2)$ in A and in B. If $\overline{A} \to \overline{C} \to \overline{B} \supset \overline{\mathfrak{q}}$ is resolvable, then $A \to C \to B \supset \mathfrak{q}$ is resolvable.

Proof Assume $H_{C|A}B \subset \mathfrak{q}$. Let $\overline{A} \to \overline{C} \to \overline{D} \to \overline{B}$ be such that $H_{\overline{C}|\overline{A}}\overline{B} \subset H_{\overline{D}|\overline{A}}\overline{B} \not\subset \overline{\mathfrak{q}}$. By (5.2.5), there exists $A \to C \to E \to B$ with $\pi^{-1}(H_{\overline{D}|\overline{A}}\overline{B}) \subset \mathrm{rad}(H_{E|A}B)$, where $\pi : B \to \overline{B}$ is the canonical map. By (5.1.14), $H_{C|A}\overline{C} \subset H_{\overline{C}|\overline{A}}$, so $H_{C|A}\overline{B} \subset H_{\overline{C}|\overline{A}}\overline{B}$, and then $H_{C|A}B \subset \pi^{-1}(H_{\overline{C}|\overline{A}}\overline{B})$. Therefore

$$H_{C|A}B \subset \pi^{-1}(H_{\overline{C}|\overline{A}}\overline{B}) \subset \pi^{-1}(H_{\overline{D}|\overline{A}}\overline{B}) \subset \mathrm{rad}(H_{E|A}B).$$

On the other hand, $\mathrm{rad}(H_{E|A}B) \not\subset \mathfrak{q}$, since $H_{\overline{D}|\overline{A}}\overline{B} \not\subset \overline{\mathfrak{q}}$ and so $\pi^{-1}(H_{\overline{D}|\overline{A}}\overline{B}) \not\subset \pi^{-1}(\overline{\mathfrak{q}}) = \mathfrak{q}$. □

Proposition 5.3.10 *Let $A \to C \to B$ be homomorphisms of noetherian rings, with C an A-algebra of finite type. Let $\mathfrak{q}$ be a prime ideal of B containing $H_{C|A}B$. Let $e \geq 1$ be an integer and $c = 8e$. Let $a_1, \ldots, a_r \in A$ be such that the image of each a_i^e in C is strictly standard, and for each i, $((a_1^c, \ldots, a_{i-1}^c) : a_i^2) = ((a_1^c, \ldots, a_{i-1}^c) : a_i)$ in A and in B (when $i = 1$ we understand $(0 : a_1^2) = (0 : a_1)$). Let $\overline{A} = A/(a_1^c, \ldots, a_r^c)$, etc.*

If $\overline{A} \to \overline{C} \to \overline{B} \supset \overline{\mathfrak{q}}$ is resolvable, then $A \to C \to B \supset \mathfrak{q}$ is resolvable.

Proof Replacing each a_i by a_i^e, we can assume that $e = 1$. If $r = 1$, the result follows from (5.3.9). Suppose that the result holds for $r - 1$, and let $A' = A/(a_1^c, \ldots, a_{r-1}^c)$, etc. By (5.2.3) the image of a_r in C' is strictly standard and so the result follows by induction applying (5.3.9) to $A' \to C' \to B' \supset \mathfrak{q}'$. □

Proposition 5.3.11 *If Theorem 5.3.7 holds whenever $ht(\mathfrak{p}) = 0$, then it holds in general.*

Proof With the notation as in (5.3.7), assume that $ht(\mathfrak{p}) > 0$. Since $\mathfrak{p}$ is minimal over $f^{-1}(H_{C|A}B)$, there exists $a \in f^{-1}(H_{C|A}B)$ such that $a \notin \mathfrak{n}$, for any prime ideal $\mathfrak{n} \subset \mathfrak{p}$ of A with $ht\,\mathfrak{n} = 0$ [AM, Proposition 1.11]. That is, $ht(a) \geq 1$, and so $ht(\mathfrak{p}/(a)) < ht\,\mathfrak{p}$. By (5.3.6), there exists $A \to C \to D \to B$ with $H_{C|A}D \subset H_{D|A}$ and the image of a in D standard. Replacing a by a^s if necessary, we can suppose that the image of a in D is strictly standard and that $(0 : a) = (0 : a^2)$ in A and in B. If $H_{D|A}B \not\subset \mathfrak{q}$, then the result holds. If $H_{D|A}B \subset \mathfrak{q}$, then $\mathfrak{q}$ is a minimal prime ideal over $H_{D|A}B$ (since it is minimal over $H_{C|A}B$), and similarly $\mathfrak{p}$ is minimal over $f^{-1}(H_{D|A}B)$. Then the result follows by induction on ht $\mathfrak{p}$, applying (5.3.9) to $A \to D \to B \supset \mathfrak{q}$. $\qquad\square$

5.4 The separable case

Proposition 5.4.1 *Let $A \to C \to B$ be homomorphisms of noetherian rings, with C an A-algebra of finite type, and $\mathfrak{q}$ a prime ideal of B minimal over $H_{C|A}B$. If $A \to C \to B_{\mathfrak{q}} \supset \mathfrak{q}B_{\mathfrak{q}}$ is resolvable, then there exists $A \to C \to D \to B$ with $H_{D|A}B \not\subset \mathfrak{q}$.*

Remark We do not claim that $H_{C|A}B \subset \mathrm{rad}(H_{D|A}B)$. See (5.4.2).

Proof By assumption, there exists an A-algebra of finite type E and a factorization $A \to C \to E \to B_{\mathfrak{q}}$ such that $H_{E|A}B_{\mathfrak{q}} \not\subset \mathfrak{q}B_{\mathfrak{q}}$, that is, $H_{E|A}B_{\mathfrak{q}} = B_{\mathfrak{q}}$. By (5.3.6.ii) we can assume that E is a smooth A-algebra. Let $E = C[X_1, \ldots, X_n]/(F_1, \ldots, F_m)$. Since n and m are finite, we can choose $s \in B - \mathfrak{q}$ such that the homomorphism $E \to B_{\mathfrak{q}}$ factors through B_s. Also, replacing s by some power, we can assume that the image of each X_i in B_s is of the form z_i/s with $z_i \in B$.

Consider the commutative diagram

$$
\begin{array}{ccc}
C[X_1, \ldots, X_n, Y] & \xrightarrow{\;\varphi\;} & E[Y] \\
{\scriptstyle \lambda}\downarrow & & \downarrow{\scriptstyle i} \\
B & \longrightarrow & B_s
\end{array}
$$

defined by $\varphi(X_i) = Yx_i$ (where $x_i \in E$ is the image of X_i), $\varphi(Y) = Y$, $\lambda(X_i) = z_i$, $\lambda(Y) = s$, $i(Y) = s$.

Since the image of $\lambda(\ker\varphi)$ in B_s is 0, there exists an integer t such that $s^t\lambda(\ker\varphi) = 0$ in B. Let $D = C[X_1, \ldots, X_n, Y]/Y^t(\ker\varphi)$. It is clear that λ induces a C-algebra homomorphism $D \to B$. If y is the class of Y in D, $D_y = E[Y]_y = E[Y, 1/Y]$ which is a smooth E-algebra, and so smooth over A. Thus $s = \lambda(Y) \in H_{D|A}B$ and then $H_{D|A}B \not\subset \mathfrak{q}$. $\square$

Proposition 5.4.2 *Let $A \to C \to B$, $\mathfrak{q}$ be as in (5.4.1). Assume moreover that $\operatorname{ht} \mathfrak{q} = 0$. If there exists a factorization $A \to C \to D \to B$ with D an A-algebra of finite type and $H_{D|A}B \not\subset \mathfrak{q}$, then there exists $A \to C \to D \to D' \to B$ with D' an A-algebra of finite type and*

$$H_{C|A}B \subset \operatorname{rad}(H_{D'|A}B) \not\subset \mathfrak{q}.$$

Proof If $H_{C|A}B$ is nilpotent, we can take $D' = D$. Assume that $H_{C|A}B$ is not nilpotent, and in particular, $\mathfrak{q}$ is not nilpotent and (0) is not a primary ideal. Let $0 = \mathfrak{a} \cap \mathfrak{b}$ be a decomposition with $\mathfrak{a}$ and $\mathfrak{b}$ ideals of B such that $\operatorname{Ass}(B/\mathfrak{a}) = \{\mathfrak{q}\}$, $\mathfrak{q} \notin \operatorname{Ass}(B/\mathfrak{b})$ [Mt, Theorems 6.6 and 6.8]. Since $\mathfrak{q}$ is not nilpotent and $\operatorname{rad} \mathfrak{a} = \mathfrak{q}$, we have $\mathfrak{b} \neq 0$.

Suppose that $\mathfrak{a} = (z_1, \ldots, z_r)$ and $\mathfrak{b} = (y_1, \ldots, y_s)$. Then if $D = C[X_1, \ldots, X_n]/(F_1, \ldots, F_m)$, let

$$D' = C[X_1, \ldots, X_n, Y_1, \ldots, Y_s, Z_1, \ldots, Z_r]/(\{Y_i F_j\}_{i,j}, \{Y_k Z_l\}_{k,l}).$$

The homomorphism $C[X_1, \ldots, X_n, Y_1, \ldots, Y_s, Z_1, \ldots, Z_r] \to B$ of algebras over $C[X_1, \ldots, X_n]$ that sends Y_i to y_i and Z_j to z_j factors through a homomorphism $\varphi \colon D' \to B$, since the image of $Y_i F_j$ is 0, and the image of $Y_k Z_l$ is $y_k z_l \in \mathfrak{b}\mathfrak{a} \subset \mathfrak{a} \cap \mathfrak{b} = 0$.

Since $\mathfrak{b} \not\subset \mathfrak{q}$, reordering $\{y_1, \ldots, y_s\}$ if necessary, we can assume $y_1 \notin \mathfrak{q}$. Let $\mathfrak{p} = \varphi^{-1}(\mathfrak{q})$. The image $\overline{Y}_1$ of Y_1 in D' does not belong to $\mathfrak{p}$ and

$$D'_{\overline{Y}_1} = (C[X_1, \ldots, X_n, Y_1, \ldots, Y_s]/(F_1, \ldots, F_m))_{\overline{Y}_1} = D[Y_1, \ldots, Y_s]_{\overline{Y}_1},$$

which is a smooth D-algebra. Therefore, $H_{D'|D} \not\subset \mathfrak{p}$. By assumption, $H_{D|A}B \not\subset \mathfrak{q}$, and so $H_{D'|A}B \not\subset \mathfrak{q}$ by (5.1.13) and [AM, Proposition 1.11.ii].

Finally, we shall see that $H_{C|A}B \subset \operatorname{rad}(H_{D'|A}B)$. Let $\mathfrak{n}$ be a prime ideal of B such that $\mathfrak{q} \not\subset \mathfrak{n}$. We have $\mathfrak{a} \not\subset \mathfrak{n}$, and so, after reordering $z_1, \ldots, z_r$ if necessary, we can assume $z_1 \notin \mathfrak{n}$. The image $\overline{Z}_1$ of Z_1 in D' does not belong to $\mathfrak{m} := \varphi^{-1}(\mathfrak{n})$ as above, and so $D'_{\overline{Z}_1}$ is isomorphic to a localization of $C[X_1, \ldots, X_n, Z_1, \ldots, Z_r]$ which is smooth over C. Then, $\mathfrak{q} \not\subset \mathfrak{n}$ implies $H_{D'|C} \not\subset \mathfrak{m} = \varphi^{-1}(\mathfrak{n})$. Therefore, if $\mathfrak{n}$ is a prime ideal of B such that $H_{C|A}B \not\subset \mathfrak{n}$, we have $\mathfrak{q} \not\subset \mathfrak{n}$, and so $H_{D'|C} \not\subset \mathfrak{m}$. On the other hand, $H_{C|A}B \not\subset \mathfrak{n}$ implies $H_{C|A}D' \not\subset \mathfrak{m}$. By (5.1.13), $H_{D'|A} \not\subset \mathfrak{m}$, and so $H_{D'|A}B \not\subset \mathfrak{n}$. Thus $H_{C|A}B \subset \operatorname{rad}(H_{D'|A}B)$. □

Lemma 5.4.3 *Let $f \colon (A, \mathfrak{m}, K) \to (B, \mathfrak{n}, L)$ be a local homomorphism of noetherian local rings. Suppose that A, B and $B \otimes_A K$ are regular and $\dim B = \dim A + \dim B \otimes_A K$. Then f is flat. If moreover $L|K$ is separable, then f is formally smooth.*

Proof By the local flatness criterion and (2.6.1), to prove the flatness of f we have to show that the homomorphism

$$\alpha\colon H_2(A, K, L) \to H_2(B, B \otimes_A K, L)$$

is surjective, and the homomorphism

$$\beta\colon H_1(A, K, L) \to H_1(B, B \otimes_A K, L)$$

is injective. The surjectivity of α is clear, since $H_2(B, B \otimes_A K, L) = 0$ by (2.5.4). On the other hand, $H_1(A, K, L) = \mathfrak{m}/\mathfrak{m}^2 \otimes_K L$, and then $\dim_K H_1(A, K, L) = \dim A$, since A is regular. Similarly, since B and $B \otimes_A K$ are regular, $\dim_K H_1(B, B \otimes_A K, L) = \dim B - \dim B \otimes_A K = \dim A$. Since β is surjective (2.6.2), it is an isomorphism.

If $L|K$ is separable, f is formally smooth by (2.6.5), (2.5.8). $\square$

Remark The assumptions of (5.4.3), though enough for our purposes, are unnecessarily strong. See, e.g., [Mt, Theorem 23.1].

Lemma 5.4.4 *Let* $A \to B \to C$ *be local homomorphisms of noetherian local rings such that B and C are flat as A-modules. Let I be a proper ideal of A. If $B/IB \to C/IC$ is flat (resp. formally smooth), then $B \to C$ is flat (resp. formally smooth).*

Proof The flat case follows easily from the local flatness criterion [Mt, Theorem 22.3, (4) $\Longrightarrow$ (1)]. The formally smooth case follows from the flat case by (2.3.5), bearing in mind that if L is the residue field of C, then

$$H_1(B, C, L) = H_1(B/IB, C/IC, L)$$

by (1.4.3). $\square$

Proof of Theorem 5.3.7 for separable residue field extension $k(\mathfrak{q})|k(\mathfrak{p})$ By (5.3.11) we can assume $\operatorname{ht}\mathfrak{p} = 0$. Note that this reduction does not affect the separability assumption. Induction on $\operatorname{ht}\mathfrak{q}$.

If $\operatorname{ht}\mathfrak{q} = 0$, by (5.4.1), (5.4.2) we can assume that A and B are artinian local rings, and the homomorphism is local. In this case, the fibre $B/\mathfrak{p}B$ is regular ($\mathfrak{p}$ is the maximal ideal of A) by (2.5.8), and so is a field, that is, $\mathfrak{p}B$ is the maximal ideal of B. Therefore B is a Cohen A-algebra (3.1.1). Since the residue field extension is separable, by (3.1.8) and its proof, we see that B is a filtered inductive limit of local subrings that are A-algebras essentially of finite type, and that are formally smooth by (3.1.2). So, as we saw in the proof that Theorems 5.3.2 and 5.3.3 are equivalent, we see now that for any $A \to C \to B$ with C an A-algebra of finite type, there exists $A \to C \to E \to B$ with E a smooth algebra essentially of finite type (2.3.8), and so by (5.3.5) there exists $A \to C \to D \to B$ with D a smooth algebra of finite type, that is $H_{D|A} = D$.

If $\operatorname{ht} \mathfrak{q} > 0$, since $\mathfrak{q}$ is minimal over $H_{C|A}B$, we have $\operatorname{rad}(H_{C|A}B)B_{\mathfrak{q}} = \mathfrak{q}B_{\mathfrak{q}}$. Since $B_{\mathfrak{q}}/\mathfrak{p}B_{\mathfrak{q}}$ is a regular local ring (2.5.8), there exists an element $\xi \in \operatorname{rad}(H_{C|A}B)$ whose image in $B_{\mathfrak{q}}/\mathfrak{p}B_{\mathfrak{q}}$ is part of a regular system of parameters. Let $A' = A[X]$, $f'\colon A' \to B$ the homomorphism of A-algebras defined by $f(X) = \xi$. Let $\mathfrak{p}' = (f')^{-1}(\mathfrak{q}) = \mathfrak{p}A' + (X)$. We have a local homomorphism $A'_{\mathfrak{p}'}/\mathfrak{p}A'_{\mathfrak{p}'} = k(\mathfrak{p})[X]_{(X)} \to B_{\mathfrak{q}}/\mathfrak{p}B_{\mathfrak{q}}$. The fibre $B_{\mathfrak{q}}/\mathfrak{p}'B_{\mathfrak{q}}$ is a regular ring since ξ is a regular parameter, and so by (5.4.3) $A'_{\mathfrak{p}'}/\mathfrak{p}A'_{\mathfrak{p}'} \to B_{\mathfrak{q}}/\mathfrak{p}B_{\mathfrak{q}}$ is formally smooth. Then applying (5.4.4) to $A_{\mathfrak{p}} \to A'_{\mathfrak{p}'} \to B_{\mathfrak{q}}$, we deduce that $A'_{\mathfrak{p}'} \to B_{\mathfrak{q}}$ is formally smooth.

Consider the factorization $A' = A[X] \to C[X] \to B$. By (5.1.14), $\xi \in \operatorname{rad}(H_{C[X]|A[X]}B)$. By (5.3.6) there exists a factorization $A[X] \to C[X] \to E \to B$ such that $H_{C[X]|A[X]}E \subset H_{E|A[X]}$ and the image of X in E is standard.

Let $a = X^s$ with s large enough such that $(0 : a^2) = (0 : a)$ in $A[X]$ and in B, and that the image of a in E is strictly standard. Let $\overline{A'} = A'/a^8A'$, etc. Since a is not a zero divisor in A', and $B_{\mathfrak{q}}$ is a flat A'-module, then a is not a zero divisor in $B_{\mathfrak{q}}$. Therefore $\operatorname{ht} \overline{\mathfrak{q}} < \operatorname{ht}(\mathfrak{q})$. If $H_{\overline{E}|\overline{A'}}B \not\subset \mathfrak{q}$ we have resolved $\overline{A'} \to \overline{E} \to \overline{B} \supset \overline{\mathfrak{q}}$. If $H_{\overline{E}|\overline{A'}}B \subset \mathfrak{q}$, since $H_{C[X]|A'}E \subset H_{E|A'}$, and by (5.1.14) we have $H_{E|A'}\overline{E} \subset H_{\overline{E}|\overline{A'}}$, we have that $\overline{\mathfrak{q}}$ is a minimal prime ideal of $H_{\overline{E}|\overline{A'}}B$. Applying then the induction assumption to $\overline{A'} \to \overline{E} \to \overline{B} \supset \overline{\mathfrak{q}}$ (note that $\operatorname{ht} \overline{\mathfrak{p}'} = 0$ since $\operatorname{ht} \mathfrak{p} = 0$, that $\overline{A'}_{\overline{\mathfrak{p}'}} \to \overline{B}_{\overline{\mathfrak{q}}}$ is formally smooth by base change and that the residue field extension remains separable, in fact, it is the same), we deduce that $\overline{A'} \to \overline{E} \to \overline{B} \supset \overline{\mathfrak{q}}$ is resolvable. Then, by (5.3.9), $A' \to E \to B \supset \mathfrak{q}$ is resolvable, that is, there exists $A' \to E \to D \to B$ with $H_{E|A'}B \subset \operatorname{rad}(H_{D|A'}B) \not\subset \mathfrak{q}$. Then $A \to C \to D \to B \supset \mathfrak{q}$ resolves $A \to C \to B \supset \mathfrak{q}$, since $H_{D|A'} \subset H_{D|A}$ by (5.1.13) (note that $A \to A'$ is smooth) and so $\operatorname{rad}(H_{C|A}B) \subset \operatorname{rad}(H_{C[X]|A'}B) \subset \operatorname{rad}(H_{E|A'}B) \subset \operatorname{rad}(H_{D|A'}B) \subset \operatorname{rad}(H_{D|A}B)$, and $\operatorname{rad}(H_{D|A}B) \not\subset \mathfrak{q}$. $\qquad \square$

5.5 Positive characteristic

Let A be an artinian local ring with residue field K of characteristic $p > 0$. By (3.2.1) there exists a Cohen ring A_0 and a local homomorphism $h\colon A_0 \to A$ inducing the identity map on the residue fields. If $C = A_0/\ker(h)$, then C is an artinian local subring of A with maximal ideal generated by p and residue field K. The construction of A_0 and so that of C is made in (3.1.7), (3.1.8). Following [An4], now we will need, in this particular case, a construction of C (of Nagata and Narita) in some sense more canonical (for each choice of a p-basis of K). This construction also works in the non-noetherian case. See also [Bo, Chapter IX, §2.2].

Definition 5.5.1 Let A be a ring, p a prime number, I an ideal of A. We define $I' = \{a^p + pb : a, b \in I\}$. If $I = A$, we have that A' is a subring of A. In general, I' is an ideal of A'. We define inductively for each integer $n \geq 0$, $A_0 = A$, $A_n = (A_{n-1})'$ for $n > 0$, and similarly I_n.

Lemma 5.5.2 *If $p \in I$, then $I_n \subset I^{n+1}$ for all $n \geq 0$.*

Proof $I_0 = I \subset I^1$. Assume $I_{n-1} \subset I^n$. Then $I_n = (I_{n-1})' \subset (I^n)' \subset (I^n)^p + pI^n \subset I^{n+1}$. $\square$

Lemma 5.5.3 *Let $(A, \mathfrak{m}, K)$ be a local ring such that the characteristic of K is $p > 0$ and $\mathfrak{m}^{r+1} = 0$ for some r. Let D be a subring of A such that $D + \mathfrak{m} = A$. Then $D_n = A_n$ for all $n \geq r$.*

Proof We have $D' + \mathfrak{m}' = A'$, since if $x \in A$, $x = d + u$ with $d \in D$, $u \in \mathfrak{m}$, then $x^p = d^p + (u^p + pv)$ with $v \in \mathfrak{m}$, and so $x^p \in D' + \mathfrak{m}'$. Similarly, $px = pd + pu \in D' + \mathfrak{m}'$. Analogously, $D_n + \mathfrak{m}_n = A_n$ for all $n \geq 0$, and since $\mathfrak{m}_r \subset \mathfrak{m}^{r+1} = 0$ by (5.5.2), we have $D_r = A_r$. $\square$

Lemma 5.5.4 *Let $(A, \mathfrak{m}, K)$ be a local ring such that the characteristic of K is $p > 0$ and $\mathfrak{m}^{r+1} = 0$ for some r. Let $\{\alpha_i\}$ be a set of elements of A such that if $\{\overline{\alpha}_i\}$ is its image in K, then $K^p(\{\overline{\alpha}_i\}) = K$. Then $A_n[\{\alpha_i\}] = A_{n+1}[\{\alpha_i\}]$ for all $n \geq r$.*

Proof The image of A_1 by the canonical map $A \to K$ is K^p, and so the image of A_{n+1} is $K^{p^{n+1}}$. Therefore the image of $A_{n+1}[\{\alpha_i\}]$ by this map is $K^{p^{n+1}}(\{\overline{\alpha}_i\}) = K$ (note that $K = K^p(\{\overline{\alpha}_i\})$, so $K^p = K^{p^2}(\{\overline{\alpha}_i^p\})$, and then $K = K^p(\{\overline{\alpha}_i\}) = K^{p^2}(\{\overline{\alpha}_i^p\})(\{\overline{\alpha}_i\}) = K^{p^2}(\{\overline{\alpha}_i\})$; in general, $K = K^{p^n}(\{\overline{\alpha}_i\}))$. Therefore $A = A_{n+1}[\{\alpha_i\}] + \mathfrak{m}$, and by (5.5.3), for all $n \geq r$ we have $A_n = (A_{n+1}[\{\alpha_i\}])_n \subset A_{n+1}[\{\alpha_i\}]$. We have seen an inclusion $A_n[\{\alpha_i\}] \subset A_{n+1}[\{\alpha_i\}]$. The reverse inclusion is obvious. $\quad\square$

Definition 5.5.5 Let $K|F$ be a field extension of characteristic $p > 0$. We say that a finite set $\{\alpha_1, \ldots, \alpha_n\}$ of elements of K is *p-independent* over F if the set $\{\alpha_1^{r_1} \cdots \alpha_n^{r_n} : 0 \leq r_i < p \text{ for all } i\}$ is linearly independent over $K^p(F)$, in other words, if $[K^p(F)(\alpha_1, \ldots, \alpha_n) : K^p(F)] = p^n$. We say that a subset B of K is *p-independent* over F if any finite subset of B is *p*-independent. We say that a subset B of K is a *p-basis* of K over F if it is *p*-independent over F and $K^p(F)(B) = K$. In this case we have $\underset{b \in B}{\otimes} {}_{K^p(F)} K^p(F)(b) = K$. If $F = F_p$ is the field of p elements, we simply say a *p*-basis of K.

For later use in Chapter 6, we point out now that if B is a ring, and A a subring containing a field of characteristic p, we also say that a finite set $\{\beta_1, \ldots, \beta_n\}$ of elements of B is *p*-independent over A if the set $\{\beta_1^{r_1} \cdots \beta_n^{r_n} : 0 \leq r_i < p \text{ for all } i\}$ is linearly independent over $A[B^p]$.

Lemma 5.5.6 *Let $K|F$ be a field extension of characteristic $p > 0$, and D a subset of K such that $K^p(F)(D) = K$. Then there exists a p-basis $B \subset D$ of K over F.*

Proof For each $G \subset D$, let $E_G = K^p(F)(G)$. Let B be a maximal subset of *p*-independent elements of D. It is enough to show that $E_B = K$, and so it is enough to see that $D \subset E_B$. Let $d \in D$. If $d \notin E_B$, then for any finite subset $H \subset B$ we have $d \notin E_H$, and $d^p \in K^p \subset E_H$. Then $[E_H(d) : E_H] = p$, and so $[E_{H \cup \{d\}} : K^p(F)] = [E_{H \cup \{d\}} : E_H][E_H : K^p(F)] = pp^{|H|} = p^{|H \cup \{d\}|}$. Thus $H \cup \{d\}$ is *p*-independent for any H, and then $B \cup \{d\}$ is *p*-independent, contradicting the maximality of B. $\quad\square$

Proposition 5.5.7 *A subset B of K is a p-basis of K over F if and only if $\{db : b \in B\}$ is a basis of the K-vector space $\Omega_{K|F}$.*

Proof In the exact sequence

$$\Omega_{K^p(F)|F} \otimes_{K^p(F)} K \xrightarrow{\alpha} \Omega_{K|F} \to \Omega_{K|K^p(F)} \to 0$$

we have $\alpha = 0$ since in $\Omega_{K|F}$ we have $da^p = pa^{p-1}da = 0$ and $d\lambda = 0$, for $a \in K$, $\lambda \in F$. Thus

$$\Omega_{K|F} = \Omega_{K|K^p(F)}.$$

Let B be a p-basis of K over F. Assume first that $B = \{b\}$ consists in one element. Then $K = K^p(F)(b)$ and so $\Omega_{K|K^p(F)}$ is generated by db. Since $K|K^p(F)$ is a non-trivial inseparable algebraic extension, by (2.4.3) we have $\Omega_{K|K^p(F)} \neq 0$ and so $\{db\}$ is a basis of this vector space.

In general, if B is a p-basis, $\bigotimes_{b \in B} {}_{K^p(F)} K^p(F)(b) = K$, then

$$\Omega_{K|K^p(F)} = \bigoplus_{b \in B} \Omega_{K^p(F)(b)|K^p(F)} \otimes_{K^p(F)(b)} K$$

and so $\{db\}_{b \in B}$ is a basis of $\Omega_{K|K^p(F)}$ by the particular case above.

Conversely, suppose that $\{db\}_{b \in B}$ is a basis of $\Omega_{K|K^p(F)}$. Consider the homomorphism

$$C := \bigotimes_{b \in B} {}_{K^p(F)} K^p(F)(b) \to K$$

and the induced exact sequence

$$\Omega_{C|K^p(F)} \otimes_C K = \bigoplus_{b \in B} \Omega_{K^p(F)(b)|K^p(F)} \otimes_{K^p(F)(b)} K$$

$$\xrightarrow{\gamma} \Omega_{K|K^p(F)} \to \Omega_{K|C} \to 0.$$

We have that γ is an isomorphism and so $\Omega_{K|C} = 0$. If L is the image of C in K (which is a field, since $L \subset K$ is an integral extension of domains and K is a field), we have $\Omega_{K|L} = \Omega_{K|C} = 0$, and since $K^p \subset L$, from (2.4.3) we deduce $L = K$. Then, by (5.5.6) there exists a p-basis $D \subset B$. We have already seen that in this case $\{db\}_{b \in D}$ is a basis of $\Omega_{K|K^p(F)}$ and so $D = B$. $\qquad\square$

Proposition 5.5.8 *Let* $(A, \mathfrak{m}, K)$ *be a local ring such that the characteristic of* K *is* $p > 0$ *and* $\mathfrak{m}^{r+1} = 0$ *for some* r. *Let* $\{\alpha_i\}$ *be a set of elements of* A *whose image* $\{\overline{\alpha}_i\}$ *in* K *is a* p-basis. *Let*

$$C = A_r[\{\alpha_i\}].$$

Then C *is an artinian local subring of* A *with maximal ideal generated by* p *and residue field* K.

If moreover A *is artinian, then* A *is a* C-module of finite type.

Proof $A' \subset A$ is an integral extension and so $C \subset A$ is also integral. Therefore C is a local ring with maximal ideal $\mathfrak{n} := \mathfrak{m} \cap C$. From the commutative diagram

$$
\begin{array}{ccc}
C & \hookrightarrow & A \\
\downarrow & & \downarrow \\
C/\mathfrak{n} & \rightarrow & K
\end{array}
$$

we deduce that the residue field $C/\mathfrak{n}$ of C is the image of C by the map $A \rightarrow K$, and this image is $K^{p^r}[\{\bar{\alpha}_i\}] = K$.

Since $p \in \mathfrak{m} \cap C = \mathfrak{n}$, we have a commutative diagram

$$
\begin{array}{ccc}
\mathbb{Z}_{(p)} & \rightarrow & C \\
\downarrow & & \downarrow \\
F_p & \rightarrow & K
\end{array}
$$

(a) We have $H_1(\mathbb{Z}_{(p)}, K, K) = (p)/(p^2) \otimes_{F_p} K$. This follows from the exact sequence

$$
0 = H_2(F_p, K, K) \rightarrow
$$
$$
H_1(\mathbb{Z}_{(p)}, F_p, K) \rightarrow H_1(\mathbb{Z}_{(p)}, K, K) \rightarrow H_1(F_p, K, K) = 0
$$

associated to $\mathbb{Z}_{(p)} \rightarrow F_p \rightarrow K$, and from (2.4.1), (2.4.5) and (1.4.1).

(b) $\Omega_{C|\mathbb{Z}_{(p)}} \otimes_C K$ is generated as K-vector space by $\{d\alpha_i \otimes 1\}$. For, consider the commutative diagram with exact row

$$
\Omega_{A_{r+1}|\mathbb{Z}_{(p)}} \otimes_{A_{r+1}} K \xrightarrow{\beta} \Omega_{C|\mathbb{Z}_{(p)}} \otimes_C K \xrightarrow{\gamma} \Omega_{C|A_{r+1}} \otimes_C K \rightarrow 0
$$

$$
\varepsilon \searrow \qquad \nearrow
$$

$$
\Omega_{A_r|\mathbb{Z}_{(p)}} \otimes_{A_r} K
$$

The elements da^p, $d(pb)$, with $a, b \in A_r$, generate $\Omega_{A_{r+1}|\mathbb{Z}_{(p)}}$ and so $\varepsilon(da^p \otimes 1) = da^p \otimes 1 = pa^{p-1}da \otimes 1 = 0$, $\varepsilon(d(pb) \otimes 1) = d(pb) \otimes 1 = pdb \otimes 1 = 0$ in $\Omega_{A_r|\mathbb{Z}_{(p)}} \otimes_{A_r} K$. Thus $\varepsilon = 0$, then $\beta = 0$, and so γ is an isomorphism. Now the result follows, since $C = A_{r+1}[\{\alpha_i\}]$, and so $\Omega_{C|A_{r+1}}$ is generated by $\{d\alpha_i\}$.

(c) The homomorphism $\Omega_{C|\mathbb{Z}_{(p)}} \otimes_C K \xrightarrow{\lambda} \Omega_{K|\mathbb{Z}_{(p)}}$ is injective (in fact, an isomorphism). This homomorphism takes the set of generators $\{d\alpha_i \otimes 1\}$ into $\{d\bar{\alpha}_i\}$, and this last is a basis of $\Omega_{K|\mathbb{Z}_{(p)}} = \Omega_{K|F_p}$ by (5.5.7), so the claim follows.

(d) $\mathfrak{n}/\mathfrak{n}^2$ is generated by the image of p. This follows from the exact sequence associated to $\mathbb{Z}_{(p)} \rightarrow C \rightarrow K$

$$
H_1(\mathbb{Z}_{(p)}, K, K) \xrightarrow{\eta} H_1(C, K, K) \rightarrow \Omega_{C|\mathbb{Z}_{(p)}} \otimes_C K \xrightarrow{\lambda} \Omega_{K|\mathbb{Z}_{(p)}} \rightarrow 0,
$$

where $H_1(\mathbb{Z}_{(p)}, K, K) = (p)/(p^2) \otimes_{F_p} K$ by (a), $H_1(C, K, K) = \mathfrak{n}/\mathfrak{n}^2$, λ is injective by (c), and so η is surjective.

(e) C is a noetherian ring. Since the maximal ideal $\mathfrak{n}$ of C is nilpotent, it is enough to show that $C/\mathfrak{n}^i$ is noetherian for all i. Any complete local ring with maximal ideal of finite type is noetherian [AM, 10.25]. So by (d), $C/\mathfrak{n}^2$ is noetherian. Assume that $C/\mathfrak{n}^i$ is noetherian and we shall see that $C/\mathfrak{n}^{i+1}$ is noetherian.

Since $\dim_K H_1(\mathbb{Z}_{(p)}, K, K) < \infty$ by (a), from the exact sequence

$$H_1(\mathbb{Z}_{(p)}, K, K) \to H_1(\mathbb{Z}/(p^i), K, K) \to H_0(\mathbb{Z}_{(p)}, \mathbb{Z}/(p^i), K) = 0$$

we deduce $\dim_K H_1(\mathbb{Z}/(p^i), K, K) < \infty$. Since $C/\mathfrak{n}^i$ is noetherian, $\dim_K H_2(C/\mathfrak{n}^i, K, K) < \infty$ by (1.4.4), and so from the exact sequence

$$H_2(C/\mathfrak{n}^i, K, K) \to H_1(\mathbb{Z}/(p^i), C/\mathfrak{n}^i, K) \to H_1(\mathbb{Z}/(p^i), K, K)$$

we deduce $\dim_K H_1(\mathbb{Z}/(p^i), C/\mathfrak{n}^i, K) < \infty$. Now using the exact sequence

$$H_1(\mathbb{Z}_{(p)}, \mathbb{Z}/(p^i), K) \to H_1(\mathbb{Z}_{(p)}, C/\mathfrak{n}^i, K) \to H_1(\mathbb{Z}/(p^i), C/\mathfrak{n}^i, K)$$

we obtain $\dim_K H_1(\mathbb{Z}_{(p)}, C/\mathfrak{n}^i, K) < \infty$.

Consider now the exact sequence

$$H_1(\mathbb{Z}_{(p)}, C/\mathfrak{n}^i, K) \xrightarrow{\eta_i} H_1(C, C/\mathfrak{n}^i, K) \to \Omega_{C|\mathbb{Z}_{(p)}} \otimes_C K$$
$$\xrightarrow{\lambda_i} \Omega_{C/\mathfrak{n}^i|\mathbb{Z}_{(p)}} \otimes_{C/\mathfrak{n}^i} K \to 0.$$

We have that λ_i is injective, since the composite homomorphism λ

$$\Omega_{C|\mathbb{Z}_{(p)}} \otimes_C K \xrightarrow{\lambda_i} \Omega_{C/\mathfrak{n}^i|\mathbb{Z}_{(p)}} \otimes_{C/\mathfrak{n}^i} K \to \Omega_{K|\mathbb{Z}_{(p)}}$$

is injective by (c). So η_i is surjective and then $\dim_K H_1(C, C/\mathfrak{n}^i, K) < \infty$, that is $\dim_K \mathfrak{n}^i/\mathfrak{n}^{i+1} < \infty$. That means that $\mathfrak{n}^i/\mathfrak{n}^{i+1}$ is a $C/\mathfrak{n}^{i+1}$-module of finite type. Since $C/\mathfrak{n}^i$ is noetherian, $\mathfrak{n}/\mathfrak{n}^i$ is a $C/\mathfrak{n}^i$-module of finite type, and so a $C/\mathfrak{n}^{i+1}$-module of finite type. Then from the exact sequence

$$0 \to \mathfrak{n}^i/\mathfrak{n}^{i+1} \to \mathfrak{n}/\mathfrak{n}^{i+1} \to \mathfrak{n}/\mathfrak{n}^i \to 0$$

we deduce that the maximal ideal $\mathfrak{n}/\mathfrak{n}^{i+1}$ of $C/\mathfrak{n}^{i+1}$ is of finite type and so $C/\mathfrak{n}^{i+1}$ is noetherian.

(f) $\mathfrak{n} = (p)$. This follows from (d) and (e) using Nakayama's lemma.

If moreover A is artinian, then each $\mathfrak{m}^i/\mathfrak{m}^{i+1}$ is a K-vector space of finite dimension, and so a C-module of finite type, since K is also the residue field of C. Since $\mathfrak{m}$ is nilpotent, we deduce that $\mathfrak{m}$ is a C-module of finite type. The C-module $A/\mathfrak{m} = K$ is also of finite type and then so is A. This could be also deduced from (3.2.2). $\qquad\qquad\square$

Theorem 5.5.9 *Suppose that $(A, \mathfrak{m}, K) \to (B, \mathfrak{n}, L)$ is a local inclusion of artinian local rings with residue fields of characteristic $p > 0$. If $\dim_L H_1(K, L, L) < \infty$, then there exists a filtered family $\mathcal{D}$ of local subrings of B and local inclusions satisfying:*

(i) *Each $D \in \mathcal{D}$ contains A, and is an A-algebra essentially of finite type.*

(ii) $B = \varinjlim D$.

(iii) *If $D_1 \subset D_2$ in $\mathcal{D}$ have maximal ideals $\mathfrak{m}_1$, $\mathfrak{m}_2$, resp., then $\mathfrak{m}_1 D_2 = \mathfrak{m}_2$, $\mathfrak{m}_1 B = \mathfrak{n}$.*

(iv) *The homomorphisms $D \to B$ are flat for all $D \in \mathcal{D}$, and so $D_1 \to D_2$ are flat for all $D_1 \subset D_2$ in $\mathcal{D}$.*

(v) *Let $D \in \mathcal{D}$, E its residue field, $x \in B$, $\bar{x}$ its image in L. Then for any large enough power q of p, there exists $D' \in \mathcal{D}$, $D \subset D'$, with residue field $E' = E(\bar{x}^q)$, such that $x^q \in D'$.*

Proof Consider the exact sequence

$$H_1(K, L, L) \to \Omega_{K|F} \otimes_K L \to \Omega_{L|F} \to \Omega_{L|K} \to 0,$$

where $F = F_p$ is the prime subfield of K. Let $\{\alpha_i\}_{i \in I} \subset A$ be a set such that its image $\{\bar{\alpha}_i\}_{i \in I}$ in K is a p-basis of K, that is, $\{d\bar{\alpha}_i\}_{i \in I}$ is a basis of the K-vector space $\Omega_{K|F}$. Choose a (finite) basis of $H_1(K, L, L)$, represent the images in $\Omega_{K|F} \otimes_K L$ of the elements of this basis as linear combination of the $d\bar{\alpha}_i \otimes 1$ and remove from $\{d\bar{\alpha}_i\}_{i \in I}$ the elements $d\bar{\alpha}_i$ appearing in this representation. We obtain a set $\{d\bar{\alpha}_j\}_{j \in J_0}$, with $J_0 \subset I$ and $I - J_0$ finite, such that its image in $\Omega_{L|F}$ is linearly independent. We can choose a basis of $\Omega_{L|F}$ of the form $\{d\bar{\alpha}_j\}_{j \in J_0} \cup \{d\bar{\beta}_j\}_{j \in J_1}$, with $\{\beta_j\}_{j \in J_1} \subset B$.

Let $B_a = B_r[\{\alpha_j\}_{j \in J_0} \cup \{\beta_j\}_{j \in J_1}]$, where $\mathfrak{n}^{r+1} = 0$, that by (5.5.8) is an artinian local subring of B with maximal ideal (p), residue field L, and B is a B_a-module of finite type. Analogously we define $A_a = A_r[\{\alpha_i\}_{i \in I}]$.

Let E be a field, $K \subset E \subset L$, with $E|K$ a field extension of finite type. Consider the exact sequence

$$\Omega_{K|F} \otimes_K E \to \Omega_{E|F} \to \Omega_{E|K} \to 0.$$

The set $\{d\bar{\alpha}_j \otimes 1\}_{j \in J_0}$ generates a subspace of finite codimension in $\Omega_{K|F} \otimes_K E$. Therefore, since $\dim_E \Omega_{E|K} < \infty$, the image of $\{d\bar{\alpha}_j \otimes 1\}_{j \in J_0}$ in $\Omega_{E|F}$ generates a subspace of finite codimension and is linearly independent (since so is in $\Omega_{L|F}$). Let $\{\gamma_h\}_{h \in H_E} \subset B_a$ be such that H_E is finite and $\{d\overline{\alpha}_j\}_{j \in J_0} \cup \{d\overline{\gamma}_h\}_{h \in H_E}$ is a basis of $\Omega_{E|F}$. We consider all possible choices of $\{\gamma_h\}_{h \in H_E}$.

Let $\pi \colon B \to L$ be the canonical homomorphism. Then $\pi^{-1}(E)$ is a local subring of B containing A and with maximal ideal $\mathfrak{n}$. Let $B_{H_E} = \pi^{-1}(E)_r[\{\alpha_j\}_{j \in J_0} \cup \{\gamma_h\}_{h \in H_E}]$. By (5.5.8), B_{H_E} is an artinian local subring of $\pi^{-1}(E)$, with maximal ideal generated by p and residue field E. Moreover, since $\{\gamma_h\}_{h \in H_E} \subset B_a$, we have $B_{H_E} \subset B_a$. Since H_E is finite, B_{H_E} is a $\pi^{-1}(E)_r[\{\alpha_j\}_{j \in J_0}]$-algebra of finite type.

Now, for each E, let $M_E = \{j \in J_1 : \overline{\beta}_j \in E\}$. The set $\{d\overline{\alpha}_j\}_{j \in J_0} \cup \{d\overline{\beta}_j\}_{j \in M_E}$ is linearly independent in $\Omega_{E|F}$ and so M_E is finite. Let $C_E = \pi^{-1}(E)_r[\{\alpha_j\}_{j \in J_0} \cup \{\beta_j\}_{j \in M_E}]$. Clearly $\{C_E\}_E$ is a filtered system and $\varinjliminf_{E} C_E = B_a$. Then, since $\pi^{-1}(E)_r[\{\alpha_j\}_{j \in J_0}] \subset C_E$, and B_{H_E} is a $\pi^{-1}(E)_r[\{\alpha_j\}_{j \in J_0}]$-algebra of finite type, there exists $E' \supset E$ such that $B_{H_E} \subset C_{E'}$. Conversely, given E', some $H_{E'}$ satisfies $M_{E'} = \{j \in J_1 : \overline{\beta}_j \in E'\} \subset \{\gamma_h\}_{h \in H_{E'}}$, and so $C_{E'} \subset B_{H_{E'}}$. So $\{B_{H_E}\}$ is also a filtered system and its limit is B_a. Since $B_{H_E} \subset B_a$ and the maximal ideals of these two rings are generated by p, by the local flatness criterion [Mt, theorem 22.3], B_a is a flat B_{H_E}-module, and so it is faithfully flat.

By (5.5.8) or (3.2.2), B is a B_a-algebra of finite type of the form $B = B_a[Y_1, \ldots, Y_n]/(F_1, \ldots, F_s)$, where the image of each Y_i in B belongs to $\mathfrak{n}$. Since $\mathfrak{n}^{r+1} = 0$, we can assume that $Y_i^{r+1} \in \{F_1, \ldots, F_s\}$ for all i. Let $\mathcal{H} = \{H_E : B_{H_E}$ contains all the coefficients of the polynomials $F_1, \ldots, F_s\}$. We have that $\{B_{H_E}\}_{H_E \in \mathcal{H}}$ is a filtered subsystem of $\{B_{H_E}\}$ with the same limit B_a. For each $H_E \in \mathcal{H}$, let $D_{H_E} = B_{H_E}[Y_1, \ldots, Y_n]/(F_1, \ldots, F_s)$, and let $\mathcal{D}' = \{D_{H_E} : H_E \in \mathcal{H}\}$. The rings D_{H_E} are local (with maximal ideal $(p, y_1, \ldots, y_n)$, where y_i is the image of Y_i in D_{H_E}, since $Y_i^{r+1} \in \{F_1, \ldots, F_s\}$ and so $y_i^{r+1} = 0$), and the homomorphisms $D_{H_E} \to D_{H_{E'}}$ are also local.

Since $B_{H_E} \to B_a$ is faithfully flat, by base change, so is $D_{H_E} \to B_a \otimes_{B_{H_E}} D_{H_E} = B$. In particular, $D_{H_E} \subset B$. The homomorphisms $D_{H_E} \subset D_{H_{E'}}$ are then faithfully flat [Mt, p. 46]. We have already proved (ii), (iii) and (iv) for $\mathcal{D}'$.

By (5.5.8), A is an A_a-module of finite type, and so an $A_r[\{\alpha_j\}_{j \in J_0}]$-algebra of finite type. Therefore, since $A_r[\{\alpha_j\}_{j \in J_0}] \subset D_{H_E}$ for any H_E, and $\{D_{H_E}\}$ is a filtered system whose limit contains A, we have $A \subset D_{H_E}$ for some $H_E \in \mathcal{H}$ and this inclusion is local since $A \subset B$ is local. We define

$$\mathcal{D} = \{D_{H_E} : H_E \in \mathcal{H} \text{ and } A \subset D_{H_E}\}.$$

Since $A \to D_{H_E}$ is a local homomorphism with residue field extension $E|K$ of finite type, we deduce that E is an A-algebra essentially of finite type and so D_{H_E} is an A-algebra essentially of finite type, since it is artinian.

Clearly we still have (ii), (iii) and (iv) for $\mathcal{D}$, and now also (i). We shall prove (v). Let $D_{H_E} \in \mathcal{D}$, $x \in B$, $\overline{x}$ its image in L. Assume first that $\overline{x}$ is algebraic over E. Replacing x by x^q, where q is a large enough power of p, we can assume that $\overline{x}$ is separable over E and $x \in B_r \subset B_a$. Let $E' = E(\overline{x})$. By (2.4.3), (2.4.5), $\Omega_{E|F} \otimes_E E' = \Omega_{E'|F}$ and so $\{d\overline{\alpha}_j\}_{j \in J_0} \cup \{d\overline{\gamma}_h\}_{h \in H_E}$ also is a basis of $\Omega_{E'|F}$. Thus, we take $H_{E'} = H_E$, $B_{H_{E'}} = \pi^{-1}(E')_r[\{\alpha_j\}_{j \in J_0} \cup \{\gamma_h\}_{h \in H_E}]$. Since $\overline{x} \in E'$, we have $x \in \pi^{-1}(E')$, and so $x^q \in \pi^{-1}(E')_r \subset B_{H_{E'}} \subset D_{H_{E'}}$ with $q = p^r$. Moreover, since $\overline{x}$ is separable over E, we have $E' = E(\overline{x}) = E(\overline{x}^q)$.

Assume now that $\overline{x}$ is transcendental over E. Replacing as above x by x^q, we can assume that $x \in B_a$. Let $E' = E(\overline{x})$. By (2.4.3), (2.4.4), $\{d\overline{\alpha}_j\}_{j \in J_0} \cup \{d\overline{\gamma}_h\}_{h \in H_E} \cup \{d\overline{x}\}$ is a basis of $\Omega_{E'|F}$. Taking $H_{E'} = H_E \cup \{w\}$ with $\gamma_w = x$, we have $x \in B_{H_{E'}} = \pi^{-1}(E')_r[\{\alpha_j\}_{j \in J_0} \cup \{\gamma_h\}_{h \in H_E} \cup \{x\}] \subset D_{H_{E'}}$. $\square$

Lemma 5.5.10 *Let $(A, \mathfrak{m}, K) \to (B, \mathfrak{n}, L)$ be a formally smooth homomorphism of noetherian local rings. Then $\dim_L H_1(K, L, L) < \infty$.*

Proof From the Jacobi–Zariski exact sequence

$$0 = H_1(A, B, L) \to H_1(A, L, L) \to H_1(B, L, L) = \mathfrak{n}/\mathfrak{n}^2$$

we deduce $\dim_L H_1(A, L, L) < \infty$, and then from the exact sequence

$$H_1(A, L, L) \to H_1(K, L, L) \to H_0(A, K, L) = 0$$

we obtain $\dim_L H_1(K, L, L) < \infty$. $\square$

Proposition 5.5.11 *Let $f \colon A \to B$ be a homomorphism of noetherian rings. Let $\mathfrak{q}$ be a prime ideal of B such that $A_{\mathfrak{p}} \to B_{\mathfrak{q}}$ is formally smooth, where $\mathfrak{p} = f^{-1}(\mathfrak{q})$. Let $K = k(\mathfrak{p})$, $L = k(\mathfrak{q})$ be their residue fields. Assume that the characteristic of K is $p > 0$. Let $n > 0$ be an integer, $\widetilde{B} = B_{\mathfrak{q}}/\mathfrak{q}^n B_{\mathfrak{q}}$, $\widetilde{A}$ the image of $A_{\mathfrak{p}}$ in $\widetilde{B}$. Let G be a finite subset of $\widetilde{B}$. Then there exists a homomorphism $f' \colon A' := A[Y_1, \ldots, Y_m] \to B$*

that extends f, and a (not necessarily filtered) family $\widetilde{\mathcal{D}}$ of local subrings of $\widetilde{B}$ satisfying:

(i) $\mathfrak{p}'B_{\mathfrak{q}} = \mathfrak{q}B_{\mathfrak{q}}$ where $\mathfrak{p}' = (f')^{-1}(\mathfrak{q})$.

(ii) $A'_{\mathfrak{p}'} \to B_{\mathfrak{q}}$ is flat.

(iii) Let $\widetilde{A}' = A'_{\mathfrak{p}'}/(\mathfrak{p}')^n A'_{\mathfrak{p}'}$. For any $\widetilde{D} \in \widetilde{\mathcal{D}}$ we have
  (a) $\widetilde{A}' \subset \widetilde{D}$;
  (b) $G \subset \widetilde{D}$;
  (c) $\widetilde{A}' \to \widetilde{D}$ is smooth essentially of finite type;
  (d) $\widetilde{D} \to \widetilde{B}$ is flat;
  (e) $\mathfrak{p}'\widetilde{D}$ is the maximal ideal of $\widetilde{D}$.

(iv) If $\widetilde{D} \in \widetilde{\mathcal{D}}$ and $\widetilde{x} \in \widetilde{B}$, then there exists a power q of p and $\widetilde{D}' \in \widetilde{\mathcal{D}}$, such that $\widetilde{D} \subset \widetilde{D}'$ and $\widetilde{x}^q \in \widetilde{D}'$.

Proof Let $b_1, \ldots, b_r$ be a set of generators of $\mathfrak{q}$. Enlarging G if necessary, we can assume that G contains the images of $b_1, \ldots, b_r$ in $\widetilde{B}$. By (5.5.10), $\dim_L H_1(K, L, L) < \infty$. Since the residue field of $\widetilde{A}$ is K, $\widetilde{A} \to \widetilde{B}$ satisfies the assumption of (5.5.9), and so there exists a family $\mathcal{D}$ of local subrings of $\widetilde{B}$ that satisfy the conditions of (5.5.9). Since G is finite, there exists some $D \in \mathcal{D}$ such that $G \subset D$. Let E be the residue field of D and let $e_1, \ldots, e_t \in D$ be such that $\{d\bar{e}_1, \ldots, d\bar{e}_t\}$ is a basis of the E-vector space $\Omega_{E|K}$ (if e is an element of D, $\widetilde{A}$, $\widetilde{B}$, etc., we denote by $\bar{e}$ its image in the residue field).

We are going to see that we can assume that (the images in $\widetilde{B}$ of) $e_1, \ldots, e_t$ are images of elements $y_1, \ldots, y_t \in B$. Let $y_1, \ldots, y_t \in B$, $s \in B - \mathfrak{q}$ be such that each e_i is the image of $y_i/s \in B_{\mathfrak{q}}$.

If $\bar{s} \in L$ is algebraic over E, replacing s by s^q, y_i/s by $s^{q-1}y_i/s^q$ with q some power of p, we can assume that $\bar{s}$ is separable over E. By (5.5.9.(v)), replacing again s by some s^q, there exists $D' \in \mathcal{D}$, $D \subset D'$, such that $\widetilde{s} \in D'$ ($\widetilde{s}$ is the image of s in $\widetilde{B}$) and the residue field of D' is $E' = E(\bar{s})$. Replacing once again, if necessary, s by s^p, we can assume that $d\bar{s} = 0$ in $\Omega_{E'|K}$ (notice that $E(\bar{s}) = E(\bar{s}^p)$). From the exact sequence

$$0 = H_1(E, E', E') \to \Omega_{E|K} \otimes_E E' \to \Omega_{E'|K} \to \Omega_{E'|E} = 0$$

(2.4.3), (2.4.5), we deduce that $\{d\bar{e}_1, \ldots, d\bar{e}_t\}$ is also a basis of $\Omega_{E'|K}$, and then, since $\bar{y}_i = \bar{s}\bar{e}_i \in E'$, we have $d\bar{y}_i = \bar{s}d\bar{e}_i + \bar{e}_i d\bar{s} = \bar{s}d\bar{e}_i$ in $\Omega_{E'|K}$. Thus $\{d\bar{y}_1, \ldots, d\bar{y}_t\}$ is a basis of $\Omega_{E'|K}$. So replacing e_i by se_i, D by D', we have that $e_1, \ldots, e_t$ are images of elements $y_1, \ldots, y_t \in B$.

If $\bar{s} \in L$ is transcendental over E, replacing again s by s^q, let $D' \in \mathcal{D}$,

$D \subset D'$, be such that $\tilde{s} \in D'$ and the residue field of D' is $E' = E(\bar{s})$ as in (5.5.9.(v)). We have an exact sequence

$$0 = H_1(E, E', E') \to \Omega_{E|K} \otimes_E E' \to \Omega_{E'|K} \to \Omega_{E'|E} \to 0$$

showing that $\{d\bar{e}_1, \ldots, d\bar{e}_t, d\bar{s}\}$ is a basis of the E'-vector space $\Omega_{E'|K}$ (2.4.3), (2.4.5). Similarly to the algebraic case, we have $d\bar{y}_i = \bar{s}d\bar{e}_i + \bar{e}_i d\bar{s}$, and so we see that $\{d\bar{y}_1, \ldots, d\bar{y}_t, d\bar{s}\}$ also is a basis of $\Omega_{E'|K}$. Therefore, replacing D by D', $\{e_1, \ldots, e_t\}$ by $\{se_1, \ldots, se_t, s\}$ we get the result.

Let $\varphi\colon A[Y_1', \ldots, Y_t'] \to B$ be the A-algebra homomorphism taking Y_i' into y_i. We show that the induced homomorphism $A[Y_1', \ldots, Y_t']_{\mathfrak{p}''} \to B_{\mathfrak{q}}$, where $\mathfrak{p}'' = \varphi^{-1}(\mathfrak{q})$, is formally smooth. By (5.4.4), it is enough to show that $H_1(K[Y_1', \ldots, Y_t'], B \otimes_A K, L) = 0$. Let $K[Y_1', \ldots, Y_t'] \to E$ be the homomorphism of K-algebras sending Y_i' into e_i. In the Jacobi–Zariski exact sequence

$$0 = H_1(K, K[Y_1', \ldots, Y_t'], L)$$
$$\to H_1(K, E, L) \to H_1(K[Y_1', \ldots, Y_t'], E, L)$$
$$\to \Omega_{K[Y_1', \ldots, Y_t']|K} \otimes_{K[Y_1', \ldots, Y_t']} L \to \Omega_{E|K} \otimes_E L$$

associated to $K \to K[Y_1', \ldots, Y_t'] \to E$, the final homomorphism is injective, because $\{d\bar{e}_1, \ldots, d\bar{e}_t\}$ is a basis of $\Omega_{E|K}$, and so

$$H_1(K, E, L) = H_1(K[Y_1', \ldots, Y_t'], E, L).$$

We thus have a commutative diagram with exact rows and columns

$$
\begin{array}{cccc}
& & H_2(E, L, L) = 0 & \\
& & \downarrow & \\
0 = H_1(A_{\mathfrak{p}}, B_{\mathfrak{q}}, L) = H_1(K, B \otimes_A K, L) & & & \\
\to H_1(K, E, L) = H_1(K[Y_1', \ldots, Y_t'], E, L) & \to & H_1(B \otimes_A K, E, L) & \\
& & \downarrow & \downarrow \\
0 = H_2(B \otimes_A K, L, L) & & & \\
\to H_1(K[Y_1', \ldots, Y_t'], B \otimes_A K, L) & & & \\
\to H_1(K[Y_1', \ldots, Y_t'], L, L) & \to & H_1(B \otimes_A K, L, L) & \\
& \downarrow & & \downarrow \\
H_1(E, L, L) & = & H_1(E, L, L) &
\end{array}
$$

showing that $H_1(K[Y_1', \ldots, Y_t'], B \otimes_A K, L) = 0$.

Let $\{y_1'', \ldots, y_s''\} \subset \{b_1, \ldots, b_r\}$ be a set whose image in $B_{\mathfrak{q}}/\mathfrak{p}'' B_{\mathfrak{q}}$ is a regular system of parameters. Let $A' = A[Y_1', \ldots, Y_t', Y_1'', \ldots, Y_s'']$ and let $A' \to B$ be the homomorphism extending φ by sending Y_i'' into y_i''.

Let $\mathfrak{p}'$ be the inverse image of $\mathfrak{q}$ in A'. The homomorphism $A'_{\mathfrak{p}'} \to B_{\mathfrak{q}}$ is flat (by (5.4.4) applied to $A_{\mathfrak{p}} \to A'_{\mathfrak{p}'} \to B_{\mathfrak{q}}$ with $I = \mathfrak{p}A_{\mathfrak{p}}$, we can reduce the flatness to the case where $A'_{\mathfrak{p}'}$ and $B_{\mathfrak{q}}$ are regular; then (5.4.3) applies). Since $\mathfrak{p}'B_{\mathfrak{q}} = \mathfrak{q}B_{\mathfrak{q}}$, by base change, the homomorphism

$$\widetilde{A}' = A'_{\mathfrak{p}'}/(\mathfrak{p}')^n A'_{\mathfrak{p}'} \to \widetilde{B} = B_{\mathfrak{q}}/\mathfrak{q}^n B_{\mathfrak{q}}$$

is flat, and since it is local, faithfully flat. In particular, it is injective. Since the images of y_i and $y_i^{''} \in \{b_1, \dots, b_r\}$ in $\widetilde{B}$ are contained in D, we have a homomorphism $\widetilde{A}' \to D$ which is injective. From the inclusion $\mathfrak{p}'D \subset \mathfrak{m}_D$ (maximal ideal of D) we deduce $\mathfrak{p}'\widetilde{B} \subset \mathfrak{m}_D\widetilde{B} = \mathfrak{q}B_{\mathfrak{q}}/\mathfrak{q}^n B_{\mathfrak{q}}$. But $\mathfrak{p}'\widetilde{B} = \mathfrak{q}B_{\mathfrak{q}}/\mathfrak{q}^n B_{\mathfrak{q}}$, and then $\mathfrak{p}'\widetilde{B} = \mathfrak{m}_D\widetilde{B}$. Now, from the faithful flatness of $D \to \widetilde{B}$, we deduce $\mathfrak{p}'D = \mathfrak{m}_D$. Using again that $\widetilde{A}' \to \widetilde{B}$ and $D \to \widetilde{B}$ are faithfully flat, we deduce that $\widetilde{A}' \to D$ is faithfully flat.

Let $\widetilde{\mathcal{D}}$ be the family of those $\widetilde{D} \in \mathcal{D}$ such that $D \subset \widetilde{D}$ and the residue field E' of $\widetilde{D}$ is separable over the residue field E of D. Properties (i), (ii), (iii)(a), (iii)(b), (iii)(d) are now clear. Condition (iii)(e) follows from (5.5.9.(iii)).

Let K' be the residue field of $\widetilde{A}'$. We have an exact sequence

$$\Omega_{K'|K} \otimes_{K'} E \to \Omega_{E|K} \to \Omega_{E|K'} \to 0$$

and, since $\Omega_{E|K}$ is generated by $d\overline{y}_1, \dots, d\overline{y}_t$, we see that the left homomorphism is surjective and then $\Omega_{E|K'} = 0$. Therefore $E|K'$ is separable (2.4.4), (2.4.5), and thus $E'|K'$ is separable. We have seen that the local homomorphism of noetherian rings $\widetilde{A}' \to \widetilde{D}$ is flat and its fibre $K' \to E'$ is a separable field extension (since $\mathfrak{p}'\widetilde{D} = \mathfrak{m}_{\widetilde{D}}$). By (2.6.5), $\widetilde{A}' \to \widetilde{D}$, is formally smooth, proving (iii)(c).

To see (iv), let $\widetilde{D} \in \widetilde{\mathcal{D}}$ with residue field E', and $\widetilde{x} \in \widetilde{B}$. By (5.5.9.(v)), there exists $\widetilde{D}'$ with $\widetilde{D} \subset \widetilde{D}'$ and residue field $E^{''} = E'(\overline{x}^q)$ such that $\widetilde{x}^q \in \widetilde{D}'$. Taking q large enough so that $\overline{x}^q$ is separable over E', we have $E^{''}|E'$ separable and so $\widetilde{D}' \in \widetilde{\mathcal{D}}$. $\qquad\square$

Lemma 5.5.12 *Let A be a noetherian local ring.*

(i) *Let $0 \to M' \to M \to M^{''} \to 0$ be an exact sequence of A-modules of finite type. Let $a_1, \dots, a_s \in A$ be a regular sequence on M' and $M^{''}$. Then it is a regular sequence in M.*

(ii) *Let $a_1, \dots, a_s, b_1, \dots, b_s \in A$ be such that $c_1, \dots, c_s$ is a regular sequence for any choice of each $c_i \in \{a_i, b_i\}$. Then, for any j, $a_1b_1, \dots, a_{j-1}b_{j-1}, c_j, \dots, c_s$ is a regular sequence for any choice of $c_j, \dots, c_s$ as before.*

Proof (i) Applying the snake lemma to the diagram

$$
\begin{array}{ccccccccc}
0 & \longrightarrow & M' & \longrightarrow & M & \longrightarrow & M'' & \longrightarrow & 0 \\
 & & \downarrow \cdot a_1 & & \downarrow \cdot a_1 & & \downarrow \cdot a_1 & & \\
0 & \longrightarrow & M' & \longrightarrow & M & \longrightarrow & M'' & \longrightarrow & 0
\end{array}
$$

we deduce that a_1 is M-regular, and we have an exact sequence

$$ 0 \to M'/a_1 M' \to M/a_1 M \to M''/a_1 M'' \to 0, $$

so we repeat the process until a_s.

  (ii) Applying (i) to the exact sequence

$$ 0 \to A/(b_1) \xrightarrow{\cdot a_1} A/(a_1 b_1) \to A/(a_1) \to 0 $$

we deduce that $a_1 b_1, c_2, \ldots, c_s$ is a regular sequence on A for any choice of each $c_i \in \{a_i, b_i\}$. We repeat with the exact sequence

$$ 0 \to A/(a_1 b_1, b_2) \xrightarrow{\cdot a_2} A/(a_1 b_1, a_2 b_2) \to A/(a_1 b_1, a_2) \to 0 $$

(the first map is injective since $a_1 b_1, a_2, \ldots$ is a regular sequence and so a_2 is regular on $A/(a_1 b_1)$), and we deduce that $a_1 b_1, a_2 b_2, c_3, \ldots, c_s$ is a regular sequence, etc. $\qquad\square$

Lemma 5.5.13 *Let A be a noetherian ring, M an A-module of finite type, $S \subset A$ a multiplicative subset. Let $a \in A$ be such that $(0 : a)_{S^{-1}M} = (0 : a^2)_{S^{-1}M}$. Then there exists $t \in S$ such that for all $n > 0$, if $s = t^n$, $(0 : as)_M = (0 : (as)^2)_M$.*

Proof We have

$$ S^{-1} \left(\frac{(0 : a^2)_M}{(0 : a)_M} \right) = \frac{(0 : a^2)_{S^{-1}M}}{(0 : a)_{S^{-1}M}} = 0. $$

So there exists $u \in S$ such that $u(0 : a^2)_M \subset (0 : a)_M$. Let N be large enough such that $(0 : u^m)_M = (0 : u^{m+1})_M$ for all $m \geq N$ and let $t = u^N$, $s = t^n$. Then $s(0 : a^2)_M \subset (0 : a)_M$ and $(0 : s)_M = (0 : s^2)_M$. If $x \in (0 : (as)^2)_M$, $x a^2 s^2 = 0$ and then $x a^2 \in (0 : s^2)_M = (0 : s)_M$. Therefore $x s \in (0 : a^2)_M$ and then $x s^2 \in (0 : a)_M$. This gives $ax \in (0 : s^2)_M = (0 : s)_M$, and so $x \in (0 : as)_M$. $\qquad\square$

Lemma 5.5.14 *Let A be a ring, $\mathfrak{p}$ a prime ideal of A, $d_1, \ldots, d_r \in \mathfrak{p}$ elements which are a regular sequence in $A_{\mathfrak{p}}$, and $c > 0$ an integer. Then* *there exist $s_1, \ldots, s_r \in A - \mathfrak{p}$ such that, for all i,*

$$(((d_1 s_1 T_1)^c, \ldots, (d_{i-1} s_{i-1} T_{i-1})^c) : d_i s_i T_i)$$
$$= (((d_1 s_1 T_1)^c, \ldots, (d_{i-1} s_{i-1} T_{i-1})^c) : (d_i s_i T_i)^2)$$

in the polynomial ring $A[T] = A[T_1, \ldots, T_r]$.

Proof By (5.5.12), for any i, $d_1^c, \ldots, d_{i-1}^c, d_i$ is a regular sequence in $A_{\mathfrak{p}}$, and again by (5.5.12), $d_1^c T_1^c, \ldots, d_{i-1}^c T_{i-1}^c, d_i T_i$ is a regular sequence in $A_{\mathfrak{p}}[T]$. Then, if $S = A - \mathfrak{p}$, and we have already $s_1, \ldots, s_{i-1}$, setting

$$M_i = A[T]/((d_1 s_1 T_1)^c, \ldots, (d_{i-1} s_{i-1} T_{i-1})^c),$$

we have

$$S^{-1} M_i = A_{\mathfrak{p}}[T]/((d_1 T_1)^c, \ldots, (d_{i-1} T_{i-1})^c)$$

and then $(0 : d_i T_i)_{S^{-1} M_i} = (0 : (d_i T_i)^2)_{S^{-1} M_i}$. Then the result follows from (5.5.13). $\qquad \square$

Proof of (5.3.7) for residue field $k(\mathfrak{q})$ of positive characteristic Let $r = \operatorname{ht} \mathfrak{q}$ and let N be large enough so that $(\eta_{B_\mathfrak{q}})^N = 0$, where $\eta_{B_\mathfrak{q}}$ is the nilradical of $B_\mathfrak{q}$, and so that $\mathfrak{q}^N B_\mathfrak{q} \subset H_{C|A} B_\mathfrak{q}$ (such an N exists, since $\mathfrak{q}$ is minimal over $H_{C|A} B$). Let $c_1, \ldots, c_M$ be a set of generators of $H_{C|A}$, and let

$$C' = C[X_1, \ldots, X_r, \{Z_{ij}\}_{i=1,\ldots,r;j=1,\ldots,M}]/(\{X_i^{2N} - \sum_j c_j Z_{ij}\}_{i=1,\ldots,r}).$$

Since C'_{c_j} is isomorphic to $C_{c_j}[X_1, \ldots, X_r, \{Z_{ih}\}_{h \neq j}]$, it follows that C'_{c_j} is a smooth $C_{c_j}[X_1, \ldots, X_r]$-algebra, and so a smooth $A[X_1, \ldots, X_r]$-algebra, since $c_j \in H_{C|A}$. That is, the image of each c_j in C' belongs to $H_{C'|A[X_1, \ldots, X_r]}$ and then $H_{C|A} C' \subset H_{C'|A[X_1, \ldots, X_r]}$. Since $X_i^{2N} \in H_{C|A} C'$ for any i, we deduce that $X_i^{2N} \in H_{C'|A[X_1, \ldots, X_r]}$ and so $X_i \in H_{C'|A[X_1, \ldots, X_r]}$.

Let $C'' = S_{C'}(I/I^2)$, where I is the kernel of a finite free presentation of the $A[X_1, \ldots, X_r]$-algebra C' (see (5.1.16)), and $\rho : C'' \to C'$ the augmentation homomorphism. As we saw in (5.1.16), $H_{C'|A[X_1, \ldots, X_r]} C'' \subset H_{C''|A[X_1, \ldots, X_r]}$ and so $H_{C|A} C'' \subset H_{C''|A[X_1, \ldots, X_r]}$. Moreover, C'' has a presentation such that the image in C'' of any element of $H_{C'|A[X_1, \ldots, X_r]}$ is standard. Let e be such that the image of any X_i^e in C'' is strictly standard over $A[X_1, \ldots, X_r]$, and let $c = 8e$. We will apply (5.5.11) with $n = N + rc$ and G containing the images of a set of generators of C over A, so that the image of C in $\widetilde{B}$ is contained in the rings $\widetilde{D}$.

Lemma 5.5.15 *Let $d_1, \ldots, d_r \in \mathfrak{p}'$ (notation of (5.5.11)), and denote by the same symbols their images in B. Then there exists $\widetilde{D} \in \widetilde{\mathcal{D}}$ and a homomorphism $C' \to B$ sending X_i into $d_i \varepsilon_i$ with $\varepsilon_i \in B - \mathfrak{q}$, and such that the image of C' in $\widetilde{B}$ is contained in $\widetilde{D}$.*

Remark We need this result after C', $d_1, \ldots, d_r$ and $\widetilde{\mathcal{D}}$ have been fixed, so this lemma does not follow immediately from (5.5.11).

Proof Let $\widetilde{D} \in \widetilde{\mathcal{D}}$. Since $\mathfrak{q}^N B_\mathfrak{q} \subset H_{C|A} B_\mathfrak{q}$, we have that $d_i^N \in H_{C|A} B_\mathfrak{q}$ for any i, and then its image $\widetilde{d}_i^N$ in $\widetilde{B}$ belongs to $H_{C|A} \widetilde{B}$. Since $\widetilde{D} \to \widetilde{B}$ is faithfully flat, $\widetilde{d}_i^N \in H_{C|A} \widetilde{B} \cap \widetilde{D} = H_{C|A} \widetilde{D}$ [Mt, Theorem 7.5]. So, for each i, there exist elements $\widetilde{z}_{ij}' \in \widetilde{D}$ $(j = 1, \ldots, M)$ such that $\widetilde{d}_i^N = \sum_{j=1}^{M} c_j \widetilde{z}_{ij}'$.

Let $z_{ij}' \in B_\mathfrak{q}$ representing $\widetilde{z}_{ij}'$. Then $d_i^N - \sum_{j=1}^{M} c_j z_{ij}' \in \mathfrak{q}^n B_\mathfrak{q} \subset \mathfrak{q}^{n-N} H_{C|A} B_\mathfrak{q}$ and then $d_i^N = \sum_{j=1}^{M} c_j (z_{ij}' + z_{ij}'')$ with $z_{ij}'' \in \mathfrak{q}^{n-N} B_\mathfrak{q}$. Therefore, $d_i^{2N} = \sum_{j=1}^{M} c_j z_{ij}$, where the image $\widetilde{z}_{ij}$ of $z_{ij} = d_i^N z_{ij}' + d_i^N z_{ij}''$ in $\widetilde{B}$ belongs to $\widetilde{D}$, since the image of $d_i^N z_{ij}'$ belongs to $\widetilde{D}$ ($\widetilde{d}_i^N, \widetilde{z}_{ij}' \in \widetilde{D}$) and $d_i^N z_{ij}'' \in \mathfrak{q}^N \mathfrak{q}^{n-N} B_\mathfrak{q} = \mathfrak{q}^n B_\mathfrak{q}$, and so the image of $d_i^N z_{ij}''$ in $\widetilde{B}$ is 0. Let $z_{ij} = w_{ij}/s_i$ with $w_{ij} \in B$ and $s_i \in B - \mathfrak{q}$. After replacing s_i by a sufficiently high power, and enlarging $\widetilde{D}$ by (5.5.11.iv), we can assume that $\widetilde{s}_i \in \widetilde{D}$, and in particular $\widetilde{w}_{ij} = \widetilde{z}_{ij} \widetilde{s}_i \in \widetilde{D}$. Let $t_i \in B - \mathfrak{q}$ be such that $t_i((s_i d_i)^{2N} - \sum_{j=1}^{M} c_j s_i^{2N-1} w_{ij}) = 0$. Again, replacing t_i by a sufficiently high power and enlarging $\widetilde{D}$, we can assume that $\widetilde{t}_i \in \widetilde{D}$.

Define $\varepsilon_i = s_i t_i$, and so the homomorphism $C' \to B$ sending X_i into $d_i \varepsilon_i$ and Z_{ij} into $s_i^{2N-1} t_i^{2N} w_{ij}$ is well defined and its image in $\widetilde{B}$ is contained in $\widetilde{D}$.   □

By (5.3.11), we can assume $\operatorname{ht} \mathfrak{p} = 0$. Since the homomorphism $A_\mathfrak{p} \to A_{\mathfrak{p}'}'$ is formally smooth, it is flat with regular fibre. Therefore, since $A_\mathfrak{p}$ is Cohen–Macaulay (it is artinian local), we deduce that $A_{\mathfrak{p}'}'$ is Cohen–Macaulay [Mt, corollary to Theorem 23.3]. Choose $d_1, \ldots, d_s \in \mathfrak{p}'$ such that its classes in $A_{\mathfrak{p}'}' \otimes_{A_\mathfrak{p}} k(\mathfrak{p}) = A_{\mathfrak{p}'}'/\mathfrak{p} A_{\mathfrak{p}'}'$ form a regular system of parameters. Since $\mathfrak{p} A_{\mathfrak{p}'}'$ is nilpotent, $d_1, \ldots, d_s$ form a system of parameters of $A_{\mathfrak{p}'}'$ and so a regular sequence.

The images of $d_1, \ldots, d_s$ in $B_\mathfrak{q}$ also form a regular sequence, since $A_{\mathfrak{p}'}' \to B_\mathfrak{q}$ is flat, and in particular, $\dim B_\mathfrak{q} \geq s = \dim A_{\mathfrak{p}'}'$. Since $d_1, \ldots, d_s$ is a system of parameters in $A_{\mathfrak{p}'}'$, $(\mathfrak{p}')^t A_{\mathfrak{p}'}' \subset (d_1, \ldots, d_s) A_{\mathfrak{p}'}'$ for some t, and so by (5.5.11.(i)), $\mathfrak{q}^t B_\mathfrak{q} \subset (d_1, \ldots, d_s) B_\mathfrak{q}$ and then $r = \dim B_\mathfrak{q} = s$. So from now on, we denote them by $d_1, \ldots, d_r$ as before. We replace now $d_1, \ldots, d_r$ by elements $d_1 s_1, \ldots, d_r s_r$ with $s_1, \ldots, s_r \in A' - \mathfrak{p}'$ satisfying (5.5.14), that we continue to denote $d_1, \ldots, d_r$; that is, we have

$$((d_1 T_1)^c, \ldots, (d_{i-1} T_{i-1})^c : d_i T_i) = ((d_1 T_1)^c, \ldots, (d_{i-1} T_{i-1})^c : (d_i T_i)^2).$$

Lemma 5.5.16 *We can enlarge* $\widetilde{D} \in \widetilde{\mathcal{D}}$ *such that there exist elements* $\delta_i \in B - \mathfrak{q}$ *satisfying:*

(i) $\widetilde{\delta}_i \in \widetilde{D}$.

(ii) There exists a homomorphism $C' \to B$ *sending* X_i *into* $d_i \delta_i$ *such that the image of* C' *in* $\widetilde{B}$ *is contained in* $\widetilde{D}$.

(iii) The elements $d_i \delta_i$ *satisfy*

$$((d_1 \delta_1)^c, \ldots, (d_{i-1} \delta_{i-1})^c : d_i \delta_i) = ((d_1 \delta_1)^c, \ldots, (d_{i-1} \delta_{i-1})^c : (d_i \delta_i)^2)$$

in B, *for all* i.

Proof Let $\varepsilon_i \in B - \mathfrak{q}$ be the elements of (5.5.15). Assume that we have already found elements $\eta_1, \ldots, \eta_{i-1} \in B - \mathfrak{q}$ such that $\delta_j := \varepsilon_j \eta_j$ $(j < i)$ satisfy condition (iii) and that by (5.5.11.(iv)) we have enlarged $\widetilde{D}$ so that the $\widetilde{\eta}_j$ lie in $\widetilde{D}$. Since $d_1 \varepsilon_1, \ldots, d_r \varepsilon_r$ is a regular sequence in $B_{\mathfrak{q}}$, by (5.5.13), there exists $\eta_i \in B - \mathfrak{q}$ such that $d_i \varepsilon_i \eta_i^q$ satisfies (iii) for all $q > 0$. Again by (5.5.11.(iv)) we can enlarge $\widetilde{D}$ so that $\widetilde{\eta}_i^q$ lies in $\widetilde{D}$ for some q. Replacing η_i by η_i^q gives the result. Note that then condition (ii) follows: (5.5.15) (see its proof) gives a map $C' \to B$ sending X_i into $d_i \varepsilon_i$, Z_{ij} into some ζ_{ij}. If we send X_i into $d_i \varepsilon_i \eta_i$ and Z_{ij} into $\eta_i^{2N} \zeta_{ij}$ we also have a map $C' \to B$. $\square$

Lemma 5.5.17 $\mathfrak{q}^n B_{\mathfrak{q}} \subset (d_1^c, \ldots, d_r^c) B_{\mathfrak{q}}$ *and* $(\mathfrak{p}')^n A'_{\mathfrak{p}'} \subset (d_1^c, \ldots, d_r^c) A'_{\mathfrak{p}'}$.

Proof We have chosen $d_1, \ldots, d_r$ so that $(d_1, \ldots, d_r) + \mathfrak{p} A'_{\mathfrak{p}'} = \mathfrak{p}' A'_{\mathfrak{p}'}$, where $\mathfrak{p} A'_{\mathfrak{p}'}$ is nilpotent. Then $\mathfrak{q} B_{\mathfrak{q}} = \mathfrak{p}' B_{\mathfrak{q}} = (d_1, \ldots, d_r) B_{\mathfrak{q}} + \eta_{B_{\mathfrak{q}}}$, where $\eta_{B_{\mathfrak{q}}}$ is the nilradical of $B_{\mathfrak{q}}$. Therefore

$$\mathfrak{q}^n B_{\mathfrak{q}} = \sum_{i=0}^{N-1} (d_1, \ldots, d_r)^{n-i} (\eta_{B_{\mathfrak{q}}})^i \subset (d_1, \ldots, d_r)^{n-(N-1)} B_{\mathfrak{q}}$$

$$= (d_1, \ldots, d_r)^{rc+1} B_{\mathfrak{q}} \subset (d_1^c, \ldots, d_r^c) B_{\mathfrak{q}}$$

(recall that $(\eta_{B_{\mathfrak{q}}})^N = 0$).

The second inclusion follows from this one applying $- \otimes_{A'_{\mathfrak{p}'}} B_{\mathfrak{q}}$, since $A'_{\mathfrak{p}'} \to B_{\mathfrak{q}}$ is faithfully flat. $\square$

Let $A'' = A'[T_1, \ldots, T_r]$, and let $A[X_1, \ldots, X_r] \to A''$ be the map sending X_i into $d_i T_i$. Let $f'': A'' \to B$ extending the map $f': A' \to B$ by sending T_i into δ_i. Let $\Gamma = A'' \otimes_{A[X_1,\ldots,X_r]} C''$ and $\Gamma \to B$ the homomorphism which over A'' is f'', and over C'' is the composite $C'' \to C'$ with the homomorphism $C' \to B$ of (5.5.16). We have $H_{C''|A[X_1,\ldots,X_r]}\Gamma \subset H_{\Gamma|A''}$ (5.1.14), and we already saw that $H_{C|A}C'' \subset H_{C''|A[X_1,\ldots,X_r]}$. Then $H_{C|A}\Gamma \subset H_{\Gamma|A''}$.

Assume that $A'' \to \Gamma \to B \supset \mathfrak{q}$ can be resolved, i.e., there exists $A'' \to \Gamma \to E \to B \supset \mathfrak{q}$ such that $H_{\Gamma|A''}B \subset \mathrm{rad}(H_{E|A''}B) \not\subset \mathfrak{q}$. Since A'' is a smooth A-algebra $H_{E|A''} \subset H_{E|A}$ and then $H_{C|A}B \subset H_{\Gamma|A''}B \subset \mathrm{rad}(H_{E|A''}B) \subset \mathrm{rad}(H_{E|A}B) \not\subset \mathfrak{q}$, and so $A \to C \to E \to B$ resolves $A \to C \to B \supset \mathfrak{q}$.

Therefore it is enough to resolve $A'' \to \Gamma \to B \supset \mathfrak{q}$. We shall apply (5.3.10) to $a_i := d_i T_i$. We have already seen that the assumption over the quotient ideals holds in A'', and by (5.5.16), it holds also in B. On the other hand, the image of X_i^e in C'' is strictly standard over $A[X_1,\ldots,X_r]$ (by the choice of e) and then, by (5.2.3), its image a_i^e in Γ is strictly standard over A''. So (5.3.10) says that it is enough to show that

$$A''/I \to \Gamma/I\Gamma \to B/IB \supset \mathfrak{q}/IB$$

is resolvable, where $I = ((d_1 T_1)^c, \ldots, (d_r T_r)^c)$.

We have $\mathrm{ht}(\mathfrak{q}/IB) = 0$ by (5.5.17) (notice that the images δ_i of T_i in $B_{\mathfrak{q}}$ are units). So by (5.4.1) and (5.4.2), it is enough to see that

$$A''/I \to \Gamma/I\Gamma \to B_{\mathfrak{q}}/IB_{\mathfrak{q}} \supset \mathfrak{q}B_{\mathfrak{q}}/IB_{\mathfrak{q}}$$

is resolvable. It is enough to find

$$A''/I \to \Gamma/I\Gamma \to E' \to B_{\mathfrak{q}}/IB_{\mathfrak{q}}$$

with E' a smooth A''/I-algebra of finite type, and by (5.3.5) it is enough to prove that there exists

$$(*) \qquad R^{-1}(A''/I) \to R^{-1}(\Gamma/I\Gamma) \to E \to B_{\mathfrak{q}}/IB_{\mathfrak{q}},$$

where $R = A'' - \mathfrak{p}''$, with $\mathfrak{p}'' := (f'')^{-1}(\mathfrak{q})$ and where E is a smooth $R^{-1}(A''/I)$-algebra of finite type.

Let $J = (d_1^c, \ldots, d_r^c) \subset A'$. Since the elements T_i belong to R, $R^{-1}I = IR^{-1}A'' = JR^{-1}A''$, and so $(*)$ can be written as

$$R^{-1}(A''/JA'') \to R^{-1}(\Gamma/J\Gamma) \to E \to B_{\mathfrak{q}}/JB_{\mathfrak{q}}.$$

Let $S = A' - \mathfrak{p}' \subset R$. By flat base change, it is enough to show that there exists

$$S^{-1}(A''/JA'') \to S^{-1}(\Gamma/J\Gamma) \to F \to B_{\mathfrak{q}}/JB_{\mathfrak{q}}$$

with F a smooth $S^{-1}(A''/JA'')$-algebra of finite type, that is,

$$A'_{\mathfrak{p}'}[T_1, \ldots, T_r]/JA'_{\mathfrak{p}'}[T_1, \ldots, T_r]$$
$$\to A'_{\mathfrak{p}'}[T_1, \ldots, T_r]/JA'_{\mathfrak{p}'}[T_1, \ldots, T_r] \otimes_{A'[T_1,\ldots,T_r]} \Gamma$$
$$\to F \to B_{\mathfrak{q}}/JB_{\mathfrak{q}}$$

and again by base change, since $(\mathfrak{p}')^n A'_{\mathfrak{p}'}[T_1, \ldots, T_r] \subset JA'_{\mathfrak{p}'}[T_1, \ldots, T_r]$ by (5.5.17), it is enough to see that there exists

$$A'_{\mathfrak{p}'}[T_1, \ldots, T_r]/(\mathfrak{p}')^n A'_{\mathfrak{p}'}[T_1, \ldots, T_r] \to$$

$$A'_{\mathfrak{p}'}[T_1, \ldots, T_r]/(\mathfrak{p}')^n A'_{\mathfrak{p}'}[T_1, \ldots, T_r] \otimes_{A'[T_1,\ldots,T_r]} \Gamma \to F \to B_{\mathfrak{q}}/(\mathfrak{p}')^n B_{\mathfrak{q}},$$

that is,

$$\widetilde{A}'[T_1, \ldots, T_r] \to \widetilde{A}'[T_1, \ldots, T_r] \otimes_{A'[T_1,\ldots,T_r]} \Gamma \to F \to \widetilde{B}$$

or equivalently

$$\widetilde{A}'[T_1, \ldots, T_r] \to \widetilde{A}'[T_1, \ldots, T_r] \otimes_{A[X_1,\ldots,X_r]} C'' \to F \to \widetilde{B}.$$

The images of $\widetilde{A}'[T_1, \ldots, T_r]$ and C'' in $\widetilde{B}$ are contained in $\widetilde{D}$ by (5.5.16), so it is enough to see that $\widetilde{A}'[T_1, \ldots, T_r] \otimes_{A[X_1,\ldots,X_r]} C'' \to \widetilde{D}$ factors through the homomorphism $\widetilde{D}[T_1, \ldots, T_r] \to \widetilde{D}$ that sends T_i to $\widetilde{\delta}_i$, since $\widetilde{A}'[T_1, \ldots, T_r] \to \widetilde{D}[T_1, \ldots, T_r]$ is a smooth homomorphism essentially of finite type (5.5.11.(iii)) and then we apply (5.3.5).

Let $C' = C[X_1, \ldots, X_r, \{Z_{ij}\}]/(\{X_i^{2N} - \sum_j c_j Z_{ij}\}_i) \to B$ be the homomorphism of (5.5.16), sending X_i into $d_i\delta_i$, and let ζ_{ij} be the image of Z_{ij} in B. The images $\widetilde{\zeta}_{ij}$ of these elements in $\widetilde{B}$ are contained in $\widetilde{D}$ by (5.5.16). The homomorphism $A[X_1, \ldots, X_r] \to A'' = A'[T_1, \ldots, T_r]$ sends X_i into d_iT_i and then we have a homomorphism of $A[X_1, \ldots, X_r]$-algebras $C[X_1, \ldots, X_r] \to \widetilde{D}[T_1, \ldots, T_r]$ sending X_i to $\widetilde{d}_iT_i$. So we define a homomorphism $C' \to \widetilde{D}[T_1, \ldots, T_r]$ of $A[X_1, \ldots, X_r]$-algebras sending Z_{ij} into $(T_i/\widetilde{d}_i)^{2N}\widetilde{\zeta}_{ij}$ (notice that $\widetilde{d}_i$ is a unit in $\widetilde{D}$, since it is a unit in $\widetilde{B}$, and the extension of the maximal ideal of $\widetilde{D}$ is the maximal ideal of $\widetilde{B}$), which makes the diagram

$$\begin{array}{ccc} C' & \to & \widetilde{D} \\ & \searrow \quad \nearrow & \\ & \widetilde{D}[T_1, \ldots, T_r] & \end{array}$$

commute. Composing with the augmentation $C'' \to C'$, we have a homomorphism $C'' \to \widetilde{D}[T_1, \ldots, T_r]$ which induces the required homomorphism

$$\widetilde{A}'[T_1, \ldots, T_r] \otimes_{A[X_1, \ldots, X_r]} C'' \to \widetilde{D}[T_1, \ldots, T_r]. \quad \square$$

5.6 The module of differentials of a regular homomorphism

Theorem 5.6.1 *Let* $f \colon A \xrightarrow{\cdot} B$ *be a regular homomorphism. Then* $\Omega_{B|A}$ *is a flat B-module.*

Proof By (5.3.2), (2.3.1) and (1.4.8), $\Omega_{B|A}$ is the direct limit of a filtered system $\{\Omega_{B_i|A}\}$ of flat B_i-modules. Since filtered direct limits commute with $\otimes$ and are exact, they preserve flatness and then we deduce the result. $\quad\square$

Proposition 5.6.2 *Let* $f \colon A \to B$ *be a homomorphism of noetherian rings. The following are equivalent:*

(i) f *is regular.*

(ii) $H_1(A, B, k(\mathfrak{q})) = 0$ *for any prime ideal* $\mathfrak{q}$ *of* B, *where* $k(\mathfrak{q}) = B_{\mathfrak{q}}/\mathfrak{q}B_{\mathfrak{q}}$.

(iii) $H_1(A, B, B) = 0$ *and* $\Omega_{B|A}$ *is a flat B-module.*

(iv) $H_1(A, B, -) = 0$.

Proof (i) $\iff$ (ii) By (2.6.5), f is regular if and only if for any prime ideal $\mathfrak{q}$ of B, the homomorphism $A_{\mathfrak{p}} \to B_{\mathfrak{q}}$ is formally smooth, where $\mathfrak{p} = f^{-1}(\mathfrak{q})$. By (2.3.5) this last condition is equivalent to $H_1(A_{\mathfrak{p}}, B_{\mathfrak{q}}, k(\mathfrak{q})) = 0$, and by (1.4.7), $H_1(A_{\mathfrak{p}}, B_{\mathfrak{q}}, k(\mathfrak{q})) = H_1(A, B, k(\mathfrak{q}))$.

(i) $\implies$ (iii) follows from (5.3.2), (5.6.1) and (1.4.8), and (iii) $\implies$ (iv) follows from (1.4.5.d). Finally, (iv) $\implies$ (ii) is trivial. $\quad\square$

6

Localization of formal smoothness

We will prove in this chapter the theorem by André [An2] of localization of formal smoothness (6.4.2). This result is also stated without proof in Matsumura's book [Mt, end of Section 32].

In the first section we prove some lemmas which will allow us to reduce the problem to simpler cases. Sections 2 and 3 are needed for the proof of the characteristic p case (6.4.1), but they also contain some results of independent interest, for instance proposition (6.3.1) which is due to Grothendieck [EGA, 0_{IV} 22.2.6]. We will finish the proof in Section 4.

We follow essentially [An3], [BR1], [Ra1], but we also use ideas from some papers in the nineties ([An5], [An6], [BM], [Ra3]) using the relative Frobenius homomorphism, in order to switch, in some sense, from H_1 to H_2, so that we can use results like (6.2.2), (6.2.4).

6.1 Preliminary reductions

Proposition 6.1.1

i) Let $f\colon (A, \mathfrak{m}, K) \to (B, \mathfrak{n}, L)$ be a flat local homomorphism of noetherian local rings. If B is regular, then A is regular. If A and $B \otimes_A K$ are regular, then so is B.

ii) Let $A \xrightarrow{f} B \xrightarrow{g} C$ be homomorphisms of noetherian rings such that g is faithfully flat and gf regular. Then f is regular.

iii) Let $A \xrightarrow{f} B \xrightarrow{g} C$ be regular homomorphisms of noetherian rings. Then gf is regular.

Proof i) This follows from the exact sequence in the proof of Proposition 4.3.7:

$$0 \to H_2(A, K, L) \to H_2(B, L, L) \to H_2(B \otimes_A K, L, L).$$

ii) Let $\mathfrak{p}$ be a prime ideal of A, $F|k(\mathfrak{p})$ a finite field extension. The ring $C \otimes_A F$ is regular, since gf is regular. The homomorphism $B \otimes_A F \to C \otimes_A F$ is faithfully flat, and so any prime ideal $\mathfrak{q}$ of $B \otimes_A F$ is the contraction of a prime ideal $\mathfrak{q}'$ of $C \otimes_A F$. Since the homomorphism $(B \otimes_A F)_\mathfrak{q} \to (C \otimes_A F)_{\mathfrak{q}'}$ is local and flat, we deduce from i) that $(B \otimes_A F)_\mathfrak{q}$ is regular.

iii) It follows from (5.6.2) and (1.4.6). $\qquad\qquad\square$

Lemma 6.1.2 *Let $f\colon (A, \mathfrak{m}, K) \to (B, \mathfrak{n}, L)$ be a formally smooth homomorphism of noetherian local rings. Then $\hat{f}\colon \hat{A} \to \hat{B}$ is formally smooth.*

Proof Consider the commutative diagram

$$
\begin{array}{ccc}
A & \xrightarrow{\;f\;} & B \\
{\scriptstyle\alpha}\downarrow & & \downarrow{\scriptstyle\beta} \\
\hat{A} & \xrightarrow{\;\hat{f}\;} & \hat{B}
\end{array}
$$

where the vertical maps are the canonical completion homomorphisms. We have a commutative diagram with exact rows (1.4.6)

$$
\begin{array}{ccccccccc}
H_2(A, L, L) & \to & H_2(B, L, L) & \to & H_1(A, B, L) & \to & H_1(A, L, L) & \to & H_1(B, L, L) \\
\downarrow {\scriptstyle =} & & \downarrow {\scriptstyle =} & & \downarrow & & \downarrow {\scriptstyle =} & & \downarrow {\scriptstyle =} \\
H_2(\hat{A}, L, L) & \to & H_2(\hat{B}, L, L) & \to & H_1(\hat{A}, \hat{B}, L) & \to & H_1(\hat{A}, L, L) & \to & H_1(\hat{B}, L, L)
\end{array}
$$

(see (4.3.3)). Therefore $H_1(A, B, L) = H_1(\hat{A}, \hat{B}, L)$, and so the result follows from (2.3.5). $\qquad\qquad\square$

Lemma 6.1.3 *Let $f\colon (A, \mathfrak{m}, K) \to (B, \mathfrak{n}, L)$ be a local homomorphism of noetherian local rings such that the canonical homomorphism $A \to \hat{A}$ is regular. If $\hat{f}\colon \hat{A} \to \hat{B}$ is regular then f is regular.*

Proof Let

$$
\begin{array}{ccc}
A & \xrightarrow{\;f\;} & B \\
{\scriptstyle\alpha}\downarrow & & \downarrow{\scriptstyle\beta} \\
\hat{A} & \xrightarrow{\;\hat{f}\;} & \hat{B}
\end{array}
$$

be as above. If $\hat{f}$ is regular then $\beta f = \hat{f}\alpha$ is regular by (6.1.1) and so, since β is faithfully flat, f is regular by (6.1.1). □

Lemma 6.1.4 *Let A be a ring, B an A-algebra, $\mathfrak{n}$ an ideal of B. Assume that B is formally smooth for the $\mathfrak{n}$-adic topology. Let E, E' be A-algebras, M, M' nilpotent ideals of E, E' respectively, $f \colon E \to E'$ a surjective homomorphism of A-algebras such that $f(M) \subset M'$. Let $\overline{f} \colon E/M \to E'/M'$ be the induced map, $p \colon E \to E/M$, $p' \colon E' \to E'/M'$ the canonical maps. Let $u \colon B \to E/M$, $v' \colon B \to E'$ be continuous homomorphisms of A-algebras (for the discrete topologies in E and E') such that $p'v' = \overline{f}u$. Then there exists a continuous homomorphism of A-algebras $v \colon B \to E$ such that $v' = fv$, $u = pv$.*

Proof Let $J = \ker f$. We have a commutative diagram of exact rows and columns

$$
\begin{array}{ccccccccc}
& & 0 & & 0 & & 0 & & \\
& & \downarrow & & \downarrow & & \downarrow & & \\
0 & \to & M \cap J & \to & M & \to & M' = (J+M)/J & & \\
& & \downarrow & & \downarrow & & \downarrow & & \\
0 & \to & J & \to & E & \xrightarrow{f} & E' = E/J & \to & 0 \\
& & \downarrow & & \downarrow p & & \downarrow p' & & \\
0 & \to & J/(M \cap J) & \to & E/M & \xrightarrow{\overline{f}} & E'/M' = E/(J+M) & \to & 0 \\
& & \downarrow & & \downarrow & & \downarrow & & \\
& & 0 & & 0 & & 0 & &
\end{array}
$$

For any $b \in B$, we have $p'v'(b) = \overline{f}u(b)$. Let $x, y \in E$ be such that $u(b) = x + M$, $v'(b) = y + J$ (we are identifying E' with E/J). Then $x - y = n + m$, $n \in J, m \in M$. Therefore we can define $v_0 \colon B \to E/(M \cap J)$ by sending b to the class of $x - m = y + n$, and so the composite canonical homomorphisms $B \xrightarrow{v_0} E/(M \cap J) \to E/M, B \xrightarrow{v_0} E/(M \cap J) \to E/J$ are u, v' resp. Since $\ker(u) \cap \ker(v') \subset \ker(v_0)$, v_0 is continuous (for the discrete topology in $E/(M \cap J)$). Since $M \cap J$ is nilpotent, by (2.2.3) there exists a continuous homomorphism of A-algebras $v \colon B \to E$ such that v_0 is the composite $B \xrightarrow{v} E \to E/(M \cap J)$. □

Lemma 6.1.5 *Let $f \colon (A, \mathfrak{m}, K) \to (B, \mathfrak{n}, L)$ be a formally smooth homomorphism of noetherian local rings. Let C be an A-algebra, I and J two ideals of C. Assume that*

  i) C is complete for the J-adic topology.

ii) The sequence $\{I^n\}_{n\geq 0}$ converges to 0, that is, each (some) power of J contains some power of I.

iii) I is closed in C.

Then, for any continuous homomorphism of A-algebras $u\colon B \to C/I$, there exists a continuous homomorphism of A-algebras $v\colon B \to C$ such that u is the composite $B \xrightarrow{v} C \to C/I$.

Proof Some power of I is contained in J, and hence $(I + J)/J$ is a nilpotent ideal of C/J. Applying (2.2.3) to the diagram

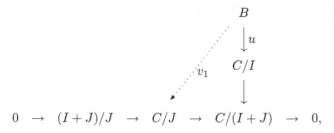

$$0 \;\to\; (I+J)/J \;\to\; C/J \;\to\; C/(I+J) \;\to\; 0,$$

we deduce the existence of a continuous homomorphism of A-algebras $v_1\colon B \to C/J$ making the diagram commute. By recurrence on n, using (6.1.4), we have a continuous homomorphism of A-algebras $v_{n+1}\colon B \to C/J^{n+1}$ making the diagram

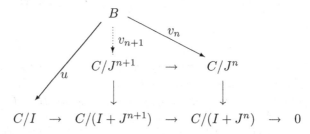

$$C/I \;\to\; C/(I+J^{n+1}) \;\to\; C/(I+J^n) \;\to\; 0$$

commute. Take $v = \varprojlim v_n\colon B \to \varprojlim C/J^n = C$. Since I is closed in C, C/I is Hausdorff and so complete. Thus $C/I = \varprojlim C/(I + J^n)$, and we obtain u as the composite of v with the canonical homomorphism $C \to C/I$. $\qquad\square$

Lemma 6.1.6 *Let*

$$\begin{array}{ccc} A' & \xrightarrow{\;g\;} & B' \\ {\scriptstyle\alpha}\downarrow & & \downarrow{\scriptstyle\beta} \\ A & \xrightarrow{\;f\;} & B \end{array}$$

be a commutative diagram of local homomorphisms of noetherian local rings, with B' complete, f formally smooth, α finite and β surjective. If g is regular, then f is regular.

Proof Let $B'' = B' \otimes_{A'} A$. Let $\mathfrak{m}$ be the maximal ideal of B'. Consider B'' with the $\mathfrak{m}B''$-adic topology and B with the topology of its maximal ideal. With this topology, B'' is complete (since B' is complete) and the canonical homomorphism $p\colon B'' \to B$ is surjective (since β is surjective) and continuous. Thus its kernel I is a closed ideal, and some power of I is contained in $\mathfrak{m}B''$ (since $B''/\mathfrak{m}B''$ is a finite $B'/\mathfrak{m}$-algebra and so artinian). We can apply (6.1.5): there exists an A-algebra homomorphism $j\colon B \to B''$ such that $pj = 1$. Therefore, for any B-module M, the identity map of $H_1(A, B, M)$ factors through $H_1(A, B' \otimes_{A'} A, M) = H_1(A', B', M) = 0$ (the first equality by (1.4.3), and the second by (5.6.2)). So $H_1(A, B, M) = 0$, and again by (5.6.2) $A \to B$ is regular. $\qquad\square$

Proposition 6.1.7 *Let $(A, (p), K)$ be a Cohen ring of positive characteristic p (3.1.10) and $(C, \mathfrak{n}, L)$ a formally smooth complete local noetherian K-algebra. Then there exists a formally smooth complete local noetherian A-algebra B such that B/pB is isomorphic to C as K-algebra.*

Proof i) Assume first that $L|K$ is separable. Then C contains a field isomorphic to L *as K-algebra*, since the canonical K-algebra epimorphism $C \to L$ has a section (2.4.6), (2.2.3). Therefore $C = L[\![X_1, \ldots, X_n]\!]$ with $n = \dim C$ as K-algebras by (3.2.2), (3.2.3), (2.5.7). Let A' be a Cohen A-algebra with residue field L (3.1.8). Then we can take $B = A'[\![X_1, \ldots, X_n]\!]$.

ii) If $L|K$ is not separable, let A_0 be a Cohen ring with residue field $F_p = \mathbb{Z}/p\mathbb{Z}$ (3.2.1). Applying i) to $(A_0, (p), F_p)$, $(C, \mathfrak{n}, L)$, we see that there exists a formally smooth complete local noetherian A_0-algebra B such that B/pB is isomorphic to C. Let $B \to C$ be the canonical homomorphism. Since $A_0 \to A$ is formally smooth (3.2.1), (3.1.2), there exists a local A_0-algebra homomorphism $f\colon A \to B$ making the diagram

$$
\begin{array}{ccc}
A & \longrightarrow & B \\
\downarrow & & \downarrow \\
K & \longrightarrow & C
\end{array}
$$

commute (6.1.5). By construction, B is a local regular ring and so p is not a zero-divisor in B. Thus, since $B/pB = C$ is flat as K-module, f is flat by the local flatness criterion [Mt, theorem 22.3] (or by (5.4.4) applied to $A_0 \to A \to B$). Then by (2.6.5) f is formally smooth. □

Proposition 6.1.8 *Let $f\colon A \to B$ be a formally smooth homomorphism of complete noetherian local rings. There exists a commutative diagram*

$$
\begin{array}{ccc}
A' & \xrightarrow{\;g\;} & B' \\
\alpha\downarrow & & \downarrow\beta \\
A & \xrightarrow{\;f\;} & B
\end{array}
$$

where A', B' are complete regular local rings, $\dim A' \le \dim A + 1$, g is formally smooth, α finite, β surjective. If A is an integral domain, we can choose A' with $\dim A' = \dim A$. If A contains a field, so does A'.

Proof i) Assume first that A contains a field. Then A contains a field isomorphic to its residue field K (3.2.4). Since $B \otimes_A K$ is a formally smooth K-algebra, $B \otimes_A K$ is a regular local ring (2.5.7), and there exists a K-algebra homomorphism $\beta_0\colon B \otimes_A K \to B$ such that the composite $B \otimes_A K \to B \to B \otimes_A K$ is the identity map ((2.2.3) or (6.1.5)). Let $\alpha\colon A' := K[\![X_1,\ldots,X_n]\!] \to A$ be a finite injective local homomorphism with $n = \dim A$ (3.2.4).

Let $\beta_1\colon B \otimes_A K[\![Y_1,\ldots,Y_m]\!] \to B$ be a surjective homomorphism extending β_0 (3.2.2) and $\beta\colon B' := B \otimes_A K[\![Y_1,\ldots,Y_m,X_1,\ldots,X_n]\!] \to B$ the homomorphism extending β_1 by $\beta(X_i) = f(\alpha(X_i))$. Finally, let $g\colon A' \to B'$ be the composite of the canonical homomorphisms

$$
\begin{aligned}
A' = K[\![X_1,\ldots,X_n]\!] &\to B \otimes_A K[\![X_1,\ldots,X_n]\!] \to B' \\
&= B \otimes_A K[\![Y_1,\ldots,Y_m,X_1,\ldots,X_n]\!],
\end{aligned}
$$

which is formally smooth because both the homomorphisms are formally smooth (the left one by base change (2.6.5), (1.4.3), (2.3.5) and the right one by (2.2.2.a), (6.1.2)).

ii) Assume now that A does not contain a field. Let $A_0 \to A$ be a local homomorphism where A_0 is a Cohen ring with residue field K (3.2.1). By (6.1.7), there exists a formally smooth complete local noetherian A_0-algebra C such that $C \otimes_{A_0} K = B \otimes_A K$. Let $\beta'\colon C \to B$ be a local homomorphism of A_0-algebras such that $C \xrightarrow{\beta'} B \to C \otimes_{A_0} K$ is the canonical homomorphism ((2.2.3) or (6.1.5)).

Let $A' := A_0[\![X_1,\ldots,X_n]\!] \to A$ be a finite homomorphism with $n = \dim A - 1$ or $n = \dim A$ (depending on whether the characteristic p of

K is part of a system of parameters of A or not; see (3.2.2), (3.2.3). In particular, $n = \dim A - 1$ if A is an integral domain, as we saw in the proof of (3.2.4.b)). Let $B' := C[\![Y_1, \ldots, Y_m, X_1, \ldots, X_n]\!]$ be as in i). The proof concludes in the same way as that of i). □

6.2 Some results on vanishing of homology

Lemma 6.2.1 *Let S be a noetherian ring, n an integer and P_* a complex of S-modules such that for any S-module of finite type N, $H_n(P_* \otimes_S N)$ is an S-module of finite type (e.g., if P_* is a complex of S-modules of finite type). If $H_n(P_* \otimes_S S/\mathfrak{m}) = 0$ for any maximal ideal $\mathfrak{m}$ of S, then $H_n(P_* \otimes_S M) = 0$ for any S-module M.*

Proof We can assume that S is local with maximal ideal $\mathfrak{m}$. Using direct limits we can assume that M is of finite type, and by the homology long exact sequence we can assume that M is cyclic. Let X be the set of ideals I of S such that $H_n(P_* \otimes_S S/I) \neq 0$. Since S is noetherian, if $X \neq \emptyset$, then X has a maximal element J (with the inclusion order). Let $a \in S - J$. Consider the exact sequence

$$0 \to S/(J:a) \xrightarrow{\cdot a} S/J \to S/(J+(a)) \to 0$$

and so the exact sequence

$$H_n(P_* \otimes_S S/(J:a)) \xrightarrow{\cdot a} H_n(P_* \otimes_S S/J) \to H_n(P_* \otimes_S S/(J+(a))).$$

Since $H_n(P_* \otimes_S S/J) \neq 0$ and $H_n(P_* \otimes_S S/(J+(a))) = 0$ by the maximality of J, we have $H_n(P_* \otimes_S S/(J:a)) \neq 0$, and so, again by the maximality of J, $J = (J:a)$. So we have a surjective homomorphism

$$H_n(P_* \otimes_S S/J) \xrightarrow{\cdot a} H_n(P_* \otimes_S S/J),$$

that is, $a H_n(P_* \otimes_S S/J) = H_n(P_* \otimes_S S/J)$. Since $H_n(P_* \otimes_S S/J)$ is an S-module of finite type, from Nakayama's lemma we deduce that a is a unit in S. Therefore the elements a of $S - J$ are units and so $J = \mathfrak{m}$, contradicting the assumption. □

Lemma 6.2.2 *Let $R \to S$ be a local homomorphism of regular local rings containing a field. Then $H_2(R, S, -) = 0$.*

Proof Let K be the prime subfield of R. Let M be an S-module, and consider the Jacobi–Zariski exact sequence

$$H_2(K, S, M) \to H_2(R, S, M) \to H_1(K, R, M).$$

We will see that $H_2(K, S, M) = 0$. By (2.5.8), the homomorphism $K \to S$ is regular, and so by (5.3.2), $S = \varinjlim S_i$, where each S_i is a smooth K-algebra of finite type. Let $\overline{K}_i$ be the residue field of S_i. By (2.5.4) we have $H_2(K, S_i, \overline{K}_i) = 0$, and so from (1.4.4), (6.2.1) we deduce that $H_2(K, S_i, -) = 0$. Finally, by (1.4.8), $H_2(K, S, -) = 0$.

Using (2.5.8) again, the homomorphism $K \to R$ is regular, and by (5.6.2) we see that $H_1(K, R, M) = 0$. From the exact sequence we deduce $H_2(R, S, M) = 0$. $\qquad\square$

Remarks

i) In the proof of (6.2.2), we used (6.2.1) to prove the following:

Let R be a noetherian ring, S an R-algebra of finite type such that $H_2(R, S, S/\mathfrak{m}) = 0$ for any maximal ideal $\mathfrak{m}$ of S. Then $H_2(R, S, -) = 0$.

The reader will notice that this fact could be proved without using (6.2.1) as follows: by (1.4.1.d), (1.4.7), (1.4.6) we can assume $R \to S$ surjective and local, and so the result would follow from (2.5.2).

However (6.2.1) will be used in (6.2.4).

ii) The assumption that the rings contain a field is not necessary, as can be seen in [An1, Supplément, Proposition 32].

Lemma 6.2.3 *Let* $f\colon A \to B$ *be a local homomorphism of complete noetherian local rings containing a field. Then for any B-module of finite type M, $H_2(A, B, M)$ is a B-module of finite type. The same result holds replacing the assumption A complete by A regular.*

Proof Let

$$
\begin{array}{ccc}
A' & \xrightarrow{\;g\;} & B' \\
\alpha\downarrow & & \downarrow\beta \\
A & \xrightarrow{\;f\;} & B
\end{array}
$$

be a commutative diagram of local homomorphisms of complete noetherian local rings containing a field with α and β surjective, A' and B'

regular; indeed, by (3.2.4) there exist surjective homomorphisms

$$A' := K[\![X_1, \ldots, X_n]\!] \xrightarrow{\alpha} A \quad \text{and} \quad B'' := L[\![Y_1, \ldots, Y_m]\!] \to B,$$

where K and L are the residue fields of A and B respectively. Take $B' := B''[\![X_1, \ldots, X_n]\!]$ and the obvious homomorphisms g and β.

In the exact sequence

$$H_2(A', B', M) \to H_2(A', B, M) \to H_2(B', B, M)$$

we have $H_2(A', B', M) = 0$ by (6.2.2), and $H_2(B', B, M)$ is a B-module of finite type by (1.4.4). Therefore $H_2(A', B, M)$ is a B-module of finite type. By (1.4.4) again, $H_1(A', A, M)$ is a B-module of finite type, and so from the exact sequence

$$H_2(A', B, M) \to H_2(A, B, M) \to H_1(A', A, M)$$

we deduce the result.

In the case that A is regular (not necessarily complete), consider the exact sequence associated to $A \to \hat{A} \to B$

$$H_2(A, \hat{A}, M) \to H_2(A, B, M) \to H_2(\hat{A}, B, M).$$

Since $H_2(A, \hat{A}, M) = 0$ by (6.2.2), and $H_2(\hat{A}, B, M)$ is a B-module of finite type by the previous case, we deduce that $H_2(A, B, M)$ is of finite type. $\qquad\square$

Lemma 6.2.4 *Let $(A, \mathfrak{m}, K) \to (B, \mathfrak{n}, L)$ be a local homomorphism of complete noetherian local rings containing a field. If $H_2(A, B, L) = 0$, then $H_2(A, B, -) = 0$. The same result holds replacing the assumption A complete by A regular.*

Proof (6.2.1) and (6.2.3). $\qquad\square$

6.3 Noetherian property of the relative Frobenius

Proposition 6.3.1 *Let $L|K$ be a field extension. The following are equivalent:*

   i) There exists a formally smooth noetherian local K-algebra A with residue field K-isomorphic to L.

   ii) $\dim_L H_1(K, L, L) < \infty$.

Proof i) $\Longrightarrow$ ii) This is a particular case of (5.5.10).

ii) $\Longrightarrow$ i) Let $V = H_1(K, L, L)$, $A_0 = S_L(V)$ the symmetric L-algebra on the L-module V and $\mathfrak{n} = \ker(S_L(V) \to L)$ the augmentation ideal. By (1.4.5.c), (2.1.4), there exists an infinitesimal extension of L over K by $H_1(K, L, L)$ associated to the identity map $id \in \mathrm{Hom}_L(H_1(K, L, L), H_1(K, L, L)) = H^1(K, L, H_1(K, L, L))$. This extension is of the form

$$0 \to H_1(K, L, L) \to A_0/\mathfrak{n}^2 \to L \to 0$$

for some K-algebra structure $f \colon K \to A_0/\mathfrak{n}^2$ on $A_0/\mathfrak{n}^2$, in general different from the structure given by the composite of the canonical homomorphisms $K \to L \to S_L(V) \to S_L(V)/\mathfrak{n}^2$ (note that the ring structure in $A_0/\mathfrak{n}^2$ is the canonical one, since any infinitesimal extension of L over K by $H_1(K, L, L)$ is trivial regarded as an extension over the prime subfield F of K:

$$H^1(K, L, H_1(K, L, L)) \to H^1(F, L, H_1(K, L, L)) = 0$$

by (2.4.5), (1.4.5.c)).

Let $A = \widehat{S_L(V)}$ be the $\mathfrak{n}$-adic completion of A_0, $\mathfrak{m} = \mathfrak{n}A$ its maximal ideal. We have a commutative diagram of F-algebras (where F is the prime subfield of K)

$$
\begin{array}{ccc}
 & & K \\
 & \overset{g}{\nearrow} & \downarrow f \\
A \overset{\quad}{\longrightarrow} & A/\mathfrak{m}^2 & = \; A_0/\mathfrak{n}^2
\end{array}
$$

where g exists by (2.2.3).

Consider A as K-algebra via g. We must show that A is formally smooth over K. Since A is a regular local ring ($A = L[\![X_1, \ldots, X_n]\!]$ with $n = \dim V$), we have $H_2(A, L, L) = 0$ by (2.5.3). On the other, inspecting the isomorphisms (1.4.5.c) and (2.1.4), we see that the canonical map $H_1(K, L, L) \to H_1(A, L, L)$ corresponds with the canonical isomorphism $H_1(K, L, L) \overset{\approx}{\to} \mathfrak{m}/\mathfrak{m}^2$ deduced from the extension

$$0 \to H_1(K, L, L) \to A_0/\mathfrak{n}^2 = A/\mathfrak{m}^2 \to L \to 0.$$

Thus the Jacobi–Zariski exact sequence

$$0 = H_2(A, L, L) \to H_1(K, A, L) \to H_1(K, L, L) \overset{\approx}{\to} H_1(A, L, L)$$

shows that $H_1(K, A, L) = 0$ and so A is formally smooth over K. $\qquad\square$

Corollary 6.3.2 *Let K be a field of characteristic $p > 0$, $(B, \mathfrak{n}, L)$ a noetherian local K-algebra such that $\dim_L H_1(K, L, L) < \infty$. Then the ring $B \otimes_K K^{1/p}$ is noetherian.*

Proof Let A be a complete formally smooth noetherian local K-algebra with residue field K-isomorphic to L (6.3.1). By (2.2.3), the homomorphism $A \to L$ induces a local homomorphism $A \to \hat{B}$, where $\hat{B}$ is the $\mathfrak{n}$-adic completion of B. By (3.2.2), replacing A by $A[\![X_1, \ldots, X_n]\!]$ if necessary, we can assume that $A \to \hat{B}$ is surjective. By (2.5.9), $A \otimes_K K^{1/p}$ is noetherian, and since $A \otimes_K K^{1/p} \to \hat{B} \otimes_K K^{1/p}$ is surjective, $\hat{B} \otimes_K K^{1/p}$ is noetherian. Since $B \to \hat{B}$ is faithfully flat, so is $B \otimes_K K^{1/p} \to \hat{B} \otimes_K K^{1/p}$, and then $B \otimes_K K^{1/p}$ is noetherian. $\qquad \square$

Lemma 6.3.3 *Let $f \colon (A, \mathfrak{m}, K) \to (B, \mathfrak{n}, L)$ be a formally smooth homomorphism of complete regular local rings containing a field of characteristic $p > 0$. Let $\phi \colon A \to A$ be the Frobenius homomorphism and $^{\phi}A$ as in (4.4.1). Then $^{\phi}A \otimes_A B$ is a noetherian local ring.*

Proof By (5.5.10), $\dim_L H_1(K, L, L) < \infty$. So, identifying K with a subfield of A (3.2.4), $^{\phi}K \otimes_K B$ is a noetherian ring (6.3.2), and so $^{\phi}K \otimes_A B$ is noetherian.

Let $^{\phi}\mathfrak{m}$ be the maximal ideal of $^{\phi}A$. From the exact sequence

$$0 \to {}^{\phi}\mathfrak{m} \to {}^{\phi}A \to {}^{\phi}K \to 0$$

we obtain an exact sequence

$$0 \to {}^{\phi}\mathfrak{m} \otimes_A B \to {}^{\phi}A \otimes_A B \to {}^{\phi}K \otimes_A B \to 0.$$

Note that $\mathrm{Spec}(^{\phi}A \otimes_A B) = \mathrm{Spec}(B)$, since we have homomorphisms $\alpha \colon B \to {}^{\phi}A \otimes_A B$, $\alpha(b) = 1 \otimes b$, $\beta \colon {}^{\phi}A \otimes_A B \to {}^{\phi}B$, $\beta(a \otimes b) = f(a)b^p$, such that $\alpha\beta$ and $\beta\alpha$ are the Frobenius homomorphisms of $^{\phi}A \otimes_A B$ and B. In particular $^{\phi}A \otimes_A B$ is local. The ideal $^{\phi}\mathfrak{m} \otimes_A B$ of $^{\phi}A \otimes_A B$ is of finite type, and the maximal ideal of $^{\phi}K \otimes_A B$ is also of finite type (since $^{\phi}K \otimes_A B$ is noetherian). Thus the maximal ideal of $^{\phi}A \otimes_A B$ is of finite type.

Let $^{\phi}A = \varinjlim A_i$, where each homomorphism $A_i \to A_j$ is local and flat, and the A-algebras A_i are of finite type (6.5.6). We have $^{\phi}A \otimes_A B = \varinjlim(A_i \otimes_A B)$. The homomorphism $A_i \to {}^{\phi}A$ is local and flat, so faithfully flat, and then the homomorphism $A_i \otimes_A B \to {}^{\phi}A \otimes_A B$ is also faithfully flat. Since $^{\phi}A \otimes_A B$ is local, so is $A_i \otimes_A B$. The maximal ideal of $^{\phi}A \otimes_A B$ is of finite type, so taking i large enough, we can assume that the image of the maximal ideal of $A_i \otimes_A B$ by the local homomorphism $A_i \otimes_A B \to {}^\phi A \otimes_A B$ generates the maximal ideal of ${}^\phi A \otimes_A B$.

So we have a filtered direct system $\{A_i \otimes_A B\}_i$ of local flat homomorphisms of noetherian local rings, such that for each i, the image of the maximal ideal of $A_i \otimes_A B$ in the limit ${}^\phi A \otimes_A B$ generates the maximal ideal of this ring. As we saw in (3.1.5) and its remark, ${}^\phi A \otimes_A B$ is noetherian.   □

6.4 End of the proof of localization of formal smoothness

Proposition 6.4.1 *Let $f\colon A \to B$ be a formally smooth homomorphism of complete regular local rings containing a field of characteristic $p > 0$. Then f is regular.*

Proof Let $\phi\colon A \to A$, $\phi\colon B \to B$ be the Frobenius homomorphisms, and ${}^\phi A$, ${}^\phi B$ as in (4.4.1). Let $\beta\colon {}^\phi A \otimes_A B \to {}^\phi B$, $\beta(x \otimes y) = f(x)y^p$.

Step 1. We shall see that ${}^\phi A \otimes_A B$ is regular. Let L be the residue field of B, and E the residue field of ${}^\phi A \otimes_A B$. This ring is noetherian and local by (6.3.3). The Jacobi–Zariski exact sequence associated to ${}^\phi A \to {}^\phi A \otimes_A B \to {}^\phi B$ (6.2.2), (1.4.3), (2.3.5)

$$0 = H_2({}^\phi A, {}^\phi B, {}^\phi L) \to H_2({}^\phi A \otimes_A B, {}^\phi B, {}^\phi L) \to H_1({}^\phi A, {}^\phi A \otimes_A B, {}^\phi L)$$
$$= H_1(A, B, {}^\phi L) = 0$$

shows that $H_2({}^\phi A \otimes_A B, {}^\phi B, {}^\phi L) = 0$. So from the Jacobi–Zariski exact sequence associated to ${}^\phi A \otimes_A B \to {}^\phi B \to {}^\phi L$ (2.5.3)

$$0 = H_2({}^\phi A \otimes_A B, {}^\phi B, {}^\phi L) \to H_2({}^\phi A \otimes_A B, {}^\phi L, {}^\phi L) \to H_2({}^\phi B, {}^\phi L, {}^\phi L) = 0$$

we deduce that $H_2({}^\phi A \otimes_A B, {}^\phi L, {}^\phi L) = 0$. Then by (2.4.11) we have

$$H_2({}^\phi A \otimes_A B, E, {}^\phi L) = 0$$

and so ${}^\phi A \otimes_A B$ is regular by (1.4.5.a) and (2.5.3).

Step 2. We have $H_2({}^\phi A \otimes_A B, {}^\phi B, -) = 0$. In fact, in the proof of step 1 we saw that $H_2({}^\phi A \otimes_A B, {}^\phi B, {}^\phi L) = 0$, and so the result follows from (6.2.4).

Step 3. Finally we shall show that f is regular. Let M be a B-module

($^\phi M$ the B-module via $\phi\colon B \to B$), and consider the Jacobi–Zariski exact sequence

$$0 = H_2(^\phi A \otimes_A B, {}^\phi B, {}^\phi M) \to H_1(^\phi A, {}^\phi A \otimes_A B, {}^\phi M) \xrightarrow{\alpha} H_1(^\phi A, {}^\phi B, {}^\phi M).$$

Then the composite homomorphism (1.4.3)

$$H_1(A, B, {}^\phi M) \xrightarrow{\approx} H_1(^\phi A, {}^\phi A \otimes_A B, {}^\phi M) \xrightarrow{\alpha} H_1(^\phi A, {}^\phi B, {}^\phi M)$$

is injective. But this homomorphism is zero by (2.4.7) or (4.4.1), and so $H_1(A, B, {}^\phi M) = 0$. In particular, if $\mathfrak{q} \in \operatorname{Spec} B$, $0 = H_1(A, B, {}^\phi k(\mathfrak{q})) = H_1(A, B, k(\mathfrak{q})) \otimes_{k(\mathfrak{q})} {}^\phi k(\mathfrak{q})$ (1.4.5.a), and so $H_1(A, B, k(\mathfrak{q})) = 0$ for all $\mathfrak{q}$. Therefore f is regular by (5.6.2). $\qquad\square$

Theorem 6.4.2 *Let* $f\colon A \to B$ *be a formally smooth homomorphism of noetherian local rings. If* A *is quasi-excellent (that is,* $A \to \widehat{A}$ *is regular), then* f *is regular.*

Proof First, we prove the following result for any $n \geq 0$:

$(*n)$ Let $f\colon A \to B$ be a formally smooth homomorphism of noetherian local rings. If A is a quasi-excellent domain of dimension n, then f is regular.

The case $n = 0$ is (2.5.8), (2.5.9). Assume we have proved $(*n)$ for $n < r$. By (6.1.2), (6.1.3), (6.1.6), (6.1.8), we can assume A and B regular and complete (note that the dimension of A has not changed).

We have to prove that for any prime ideal $\mathfrak{p}$ of A, the fiber in 0 of the homomorphism $A/\mathfrak{p} \to B/\mathfrak{p}B$ is geometrically regular. The homomorphism $A/\mathfrak{p} \to B/\mathfrak{p}B$ is formally smooth by (2.3.5), (1.4.3), (2.6.5). So by $(*n)$ for $n < r$, it is enough to consider the case $\mathfrak{p} = 0$. If the field of fractions Q of A is of characteristic zero, since B is regular, and so $B \otimes_A Q = S^{-1}B$ with $S = A - \{0\}$ is also regular, the result follows from (2.5.8). If the characteristic of Q is $p > 0$, then A contains the field $\mathbb{Z}/p\mathbb{Z}$ and this case was proved in (6.4.1).

So we have proved the result whenever A is a domain. Since we can assume A regular local by (6.1.2), (6.1.3), (6.1.6), (6.1.8), the proof is complete. $\qquad\square$

6.5 Appendix: Power series

We prove here a theorem of André [An3, théorème 23] used in (6.3.3).

Definition 6.5.1 Let $m \geq 1$ be an integer. We define an order relation in $I = \{r = (r_1, \ldots, r_m) : r_1, \ldots, r_m$ are integers $\geq 0\}$ by:

$r < s$ if and only if one of the following conditions holds:

i) $\sum_{i=1}^{m} r_i < \sum_{i=1}^{m} s_i$.

ii) $\sum_{i=1}^{m} r_i = \sum_{i=1}^{m} s_i$ and there exists an integer $1 \leq j \leq m$ such that $r_i = s_i$ for all $i > j$ and $r_j < s_j$.

It is a total order.

Lemma 6.5.2 *Let $\{r(n)\}_n \geq 0$ be an infinite sequence in I. Then there exists $0 \leq v < u$ such that $r(v)_i \leq r(u)_i$ for all $i = 1, \ldots, m$ (we will say that $r(v)$ is below $r(u)$).*

Proof We omit the proof of this lemma since it is easier to think by oneself than to read (hint: think first the case $m = 2$). $\qquad \square$

Let $L|K$ be a field extension of characteristic p with $L^p \subset K$. Let $T = (T_1, \ldots, T_m)$ be variables, $A := K[\![T]\!] \subset B := L[\![T]\!]$. We have $B^p \subset A$. If $t \in I$, we denote $\tau_t(\sum_{r \in I} \alpha_r T^r) = \sum_{r \leq t} \alpha_r T^r$, where if $r = (r_1, \ldots, r_m)$ then $T^r = T_1^{r_1} \cdots T_m^{r_m}$.

Definition 6.5.3 Let $b_1, \ldots, b_n \in B$, let $b_i(0)$ be the degree 0 term of b_i. We say that $b_1, \ldots, b_n$ are *perfectly p-independent* over A if the elements $b_1(0), \ldots, b_n(0) \in L$ are p-independent over K (5.5.5).

If R is a ring containing a field of characteristic p, let $R\{n\}$ be the submodule of the R-module $R[\underline{X}] = R[X_1, \ldots, X_n]$ consisting in the polynomials of the form

$$P(\underline{X}) = \sum_{0 \leq v_i < p} \alpha_{v_1, \ldots, v_n} X_1^{v_1} \cdots X_n^{v_n}.$$

In the case $R = A$, each coefficient $\alpha_{v_1, \ldots, v_n}$ is of the form

$$\alpha_{v_1, \ldots, v_n} = \sum_{r \in I} \lambda_{v_1, \ldots, v_n, r} T^r \text{ with } \lambda_{v_1, \ldots, v_n, r} \in K.$$

For each $r \in I$, let $f_r(\underline{X}) = \sum_{0 \leq v_i < p} \lambda_{v_1, \ldots, v_n, r} X_1^{v_1} \cdots X_n^{v_n} \in K\{n\}$. Then we can write

$$P(\underline{X}) = \sum_{r \in I} f_r(\underline{X}) T^r \in K\{n\}[\![T]\!]$$

since the number of monomials $X_1^{v_1} \cdots X_n^{v_n}$ is finite.

Remark 6.5.4 If $b_1, \ldots, b_n \in B$ are perfectly p-independent over A, then they are p-independent over A in the sense of (5.5.5). In fact, if $b_1, \ldots, b_n \in B$ are not p-independent over A, there exists $0 \neq P(\underline{X}) \in A\{n\}$ as before such that $P(b_1, \ldots, b_n) = 0$. Let r be the least index such that $f_r(\underline{X}) \neq 0$ in $K\{n\}$. Then $f_r(b_1(0), \ldots, b_n(0)) = 0$ showing that $b_1(0), \ldots, b_n(0)$ are not p-independent over K.

Proposition 6.5.5 *Let u be an element of B and $b_1, \ldots, b_\alpha$ a finite number of elements of B perfectly p-independent over A. Then there exists a finite number of elements $c_1, \ldots, c_\beta$ of B such that $b_1, \ldots, b_\alpha, c_1, \ldots, c_\beta$ are perfectly p-independent over A and u belongs to the A-subalgebra generated by them.*

Proof Let $\bar{b}_i = b_i(0), \bar{c}_j = c_j(0)$. Let t be an element of I and let $\int_t$ be the following situation (in which β depends on t)

i) the elements $\bar{b}_1, \ldots, \bar{b}_\alpha, \bar{c}_1, \ldots, \bar{c}_\beta$ of L are p-independent over K.
ii) there exists a polynomial

$$P(X_1, \ldots, X_\alpha, Y_1, \ldots, Y_\beta) = \sum_{r \in I} f_r(X_1, \ldots, X_\alpha, Y_1, \ldots, Y_\beta)T^r$$

of $K\{\alpha + \beta\}[[T]]$ such that $\tau_t(P(b_1, \ldots, b_\alpha, c_1, \ldots, c_\beta)) = \tau_t(u)$.
iii) there exist elements $r_1 < \cdots < r_\beta \leq t$ of I, none of them below another, such that for all $i = 1, \ldots, \beta$, the polynomial $\tau_{r_i}(P(X_1, \ldots, X_\alpha, Y_1, \ldots, Y_\beta) - T^{r_i}Y_i)$ is in fact a polynomial in the variables $X_1, \ldots, X_\alpha, Y_1, \ldots, Y_{i-1}$.

Let s be the next element to t. We are going to see how to obtain $\int_s$ from $\int_t$.
Let $\gamma \in L$ be such that $\tau_s(P(b_1, \ldots, b_\alpha, c_1, \ldots, c_\beta) - \gamma T^s) = \tau_s(u)$.

Case a. $\gamma \in K(\bar{b}_1, \ldots, \bar{b}_\alpha, \bar{c}_1, \ldots, \bar{c}_\beta)$. Then there exists

$$f(X_1, \ldots, X_\alpha, Y_1, \ldots, Y_\beta) \in K\{\alpha + \beta\}$$

with $f(\bar{b}_1, \ldots, \bar{b}_\alpha, \bar{c}_1, \ldots, \bar{c}_\beta) = \gamma$. Consider the elements $c_1, \ldots, c_\beta$ as in $\int_t$, replace $P(X_1, \ldots, X_\alpha, Y_1, \ldots, Y_\beta)$ by $P(X_1, \ldots, X_\alpha, Y_1, \ldots, Y_\beta) + T^s f(X_1, \ldots, X_\alpha, Y_1, \ldots, Y_\beta)$, and $r_1 < \cdots < r_\beta$ as in $\int_t$. We obtain $\int_s$.

Case b. $\gamma \notin K(\bar{b}_1, \ldots, \bar{b}_\alpha, \bar{c}_1, \ldots, \bar{c}_\beta)$ and at least one r_i is below s. Let r_λ be the first of the r_i that is below s. We have $s - r_\lambda \in I$. We keep the same polynomial P, the r_i and the c_j with $j \neq \lambda$, and we replace c_λ by $c_\lambda + \gamma T^{s-r_\lambda}$. Since $r_\lambda \leq t < s$, we have $s - r_\lambda > 0$ and so $\bar{c}_\lambda = c_\lambda(0)$ does not change, so the condition i) of $\int_s$ holds. Condition iii) holds trivially, and ii) follows since by iii),

$$P(X_1, \ldots, X_\alpha, Y_1, \ldots, Y_\lambda + W, \ldots, Y_\beta) - P(X_1, \ldots, X_\alpha, Y_1, \ldots, Y_\beta)$$
$$= T^{r_\lambda} W + \sum_{r > r_\lambda} g_r(X_1, \ldots, X_\alpha, Y_1, \ldots, Y_\beta, W) W T^r$$

with $g_r \in K\{\alpha + \beta + 1\}$. Therefore

$$\tau_s(P(b_1, \ldots, b_\alpha, c_1, \ldots, c_\lambda + \gamma T^{s-r_\lambda}, \ldots, c_\beta))$$
$$= \tau_s(P(b_1, \ldots, b_\alpha, c_1, \ldots, c_\beta) + \gamma T^s) = \tau_s(u).$$

Case c. $\gamma \notin K(\bar{b}_1, \ldots, \bar{b}_\alpha, \bar{c}_1, \ldots, \bar{c}_\beta)$ and there is no r_i below s. In this case we replace β by $\beta + 1$, we keep $c_1, \ldots, c_\beta$ but adding $c_{\beta+1} = \gamma$, we replace $P(X_1, \ldots, X_\alpha, Y_1, \ldots, Y_\beta)$ by $P(X_1, \ldots, X_\alpha, Y_1, \ldots, Y_\beta, Y_{\beta+1}) = P(X_1, \ldots, X_\alpha, Y_1, \ldots, Y_\beta) + T^s Y_{\beta+1}$, and we keep $r_1 < \cdots < r_\beta$ but adding $r_{\beta+1} = s$. We obtain $\int_s$.

Notice that in cases a and b we change neither β nor the elements r_i. In case c, we replace β by $\beta + 1$ and we add an element $r_{\beta+1} = s$. So by (6.5.2), this last case occurs only a finite number of times. On the other hand, the polynomial P in the step from $\int_t$ to $\int_s$ changes only in degree s and the c_i change only in degree $s - r_\lambda$. In particular, once we have done a finite number of steps large enough so that β does not change, the elements c_i from $\int_t$ to $\int_s$ do not change in degrees $< s - r_\beta$, and in particular this does not affect their property of being perfectly p-independent.

We can then prove the proposition by induction. We start with $\beta = 0$, $P(X_1, \ldots, X_\alpha) = 0$, and we take $\gamma = \bar{u} = u(0)$. We take limits, obtaining a polynomial $P(X_1, \ldots, X_\alpha, Y_1, \ldots, Y_\beta) \in A\{\alpha + \beta\}$ and elements $c_1, \ldots, c_\beta$ such that $b_1, \ldots, b_\alpha, c_1, \ldots, c_\beta$ are perfectly p-independent over A and such that $P(b_1, \ldots, b_\alpha, c_1, \ldots, c_\beta) = u$. $\qquad\square$

Theorem 6.5.6 *Let B be a complete noetherian local ring containing a field of characteristic $p > 0$. Then we have an isomorphism of B-algebras $^\phi B = \varinjlim B_i$ where each B_i is a local B-algebra of finite type and the homomorphisms $B_i \to B_j$ are local and flat.*

Proof First we will see that if a ring B has the property of the theorem and N is an ideal of B, then $C = B/N$ has the same property. Let $N = (x_1, \ldots, x_n)$. Let $^\phi B = \varinjlim B_i$ be as in the statement, and let j_0 be such that there exist elements $y_1, \ldots, y_n \in B_{j_0}$ whose images in $^\phi B$ are $x_1, \ldots, x_n$. Let $J = \{j : j_0 \le j\}$. We have $^\phi B = \varinjlim_{j \in J} B_j$. For each $j \in J$, let $C_j = B_j/(y_1, \ldots, y_n)B_j$. The homomorphisms $C_j \to C_t$ are obtained from $B_j \to B_t$ by base change, and so they are flat. On the other hand, taking limits in the exact sequences

$$0 \to (y_1, \ldots, y_n)B_j \to B_j \to C_j \to 0$$

we see that $^\phi C = \varinjlim_{j \in J} C_j$.

Therefore, by (3.2.4) we can suppose $B = L[\![T]\!]$ where L is a field. Since B is reduced, we can identify the homomorphism $\phi\colon B \to \ ^\phi B$ to the inclusion $B^p \subset B$. Set $K = L^p$ and we apply (6.5.5). We take as $\{B_i\}$ the $A = K[\![T]\!]$-subalgebras of B generated by a finite set $\{b_{i_1}, \ldots, b_{i_{\alpha_i}}\}$ of elements of B perfectly p-independent over A, and when $\{b_{i_1}, \ldots, b_{i_{\alpha_i}}\} \subset \{b_{j_1}, \ldots, b_{j_{\alpha_j}}\}$ we define a homomorphism $B_i \to B_j$ as the inclusion map, which is local and flat (by (6.5.4)). Since each B_i is an A-algebra of finite type, it is also a $B^p = K[\![T^p]\!]$-algebra of finite type. So the result follows. $\qquad\square$

Appendix: Some exact sequences

In the text we have used some exact sequences, which we have deduced from spectral sequences. In this appendix we give an elementary proof of those exact sequences, avoiding spectral sequences, so that the text could be read without the use of spectral sequences.

Proposition 1.4.5.d. *Let $A \to B \to C$ be ring homomorphisms and M a C-module. Then there exist exact sequences*

$$\mathrm{Tor}_2^C(\Omega_{B|A} \otimes_B C, M) \to$$
$$H_1(A, B, C) \otimes_C M \to H_1(A, B, M) \to \mathrm{Tor}_1^C(\Omega_{B|A} \otimes_B C, M) \to 0$$

and

$$0 \to \mathrm{Ext}_C^1(\Omega_{B|A} \otimes_B C, M) \to H^1(A, B, M) \to \mathrm{Hom}_C(H_1(A, B, C), M)$$
$$\to \mathrm{Ext}_C^2(\Omega_{B|A} \otimes_B C, M).$$

Proof Let $p\colon R \to B$ be a surjective A-algebra homomorphism, with R a polynomial ring over A. We can write the Jacobi–Zariski exact sequence associated to $A \to R \to B$

$$H_1(A, R, M) \to H_1(A, B, M) \to H_1(R, B, M) \to H_0(A, R, M)$$
$$\to H_0(A, B, M) \to H_0(R, B, M)$$

as $(*)$:

$$0 \to H_1(A, B, M) \to I/I^2 \otimes_B M \to \Omega_{R|A} \otimes_R M \to \Omega_{B|A} \otimes_B M \to 0,$$

where $I = \ker p$ (1.4.1).

Taking $M = C$, we obtain exact sequences (defining W)

$$0 \to H_1(A, B, C) \to I/I^2 \otimes_B C \to W \to 0$$
$$0 \to W \to \Omega_{R|A} \otimes_R C \to \Omega_{B|A} \otimes_B C \to 0$$

and hence exact sequences $(**)$

$$\mathrm{Tor}_2^C(\Omega_{B|A} \otimes_B C, M) = \mathrm{Tor}_1^C(W, M) \to H_1(A, B, C) \otimes_C M$$
$$\to I/I^2 \otimes_B M \to W \otimes_C M \to 0$$

and $(***)$

$$0 \to \mathrm{Tor}_1^C(\Omega_{B|A} \otimes_B C, M) \to W \otimes_C M \to \Omega_{R|A} \otimes_R M$$
$$\to \Omega_{B|A} \otimes_B M \to 0.$$

We can put $(*)$, $(**)$ and $(***)$ together into a commutative diagram with exact rows and columns:

$$\mathrm{Tor}_2^C(\Omega_{B|A} \otimes_B C, M)$$
$$\downarrow$$
$$H_1(A, B, C) \otimes_C M$$
$$\downarrow$$

$0 \to \quad H_1(A, B, M) \quad \to I/I^2 \otimes_B M \to \Omega_{R|A} \otimes_R M \to \Omega_{B|A} \otimes_B M \to 0$

$$\downarrow \qquad\qquad \downarrow \qquad\qquad \| \qquad\qquad \|$$

$0 \to \mathrm{Tor}_1^C(\Omega_{B|A} \otimes_B C, M) \to \quad W \otimes_C M \quad \to \Omega_{R|A} \otimes_R M \to \Omega_{B|A} \otimes_B M \to 0$

$$\downarrow$$
$$0$$

and so we deduce an exact sequence

$$\mathrm{Tor}_2^C(\Omega_{B|A} \otimes_B C, M) \to H_1(A, B, C) \otimes_C M \to H_1(A, B, M)$$
$$\to \mathrm{Tor}_1^C(\Omega_{B|A} \otimes_B C, M) \to 0$$

The proof for cohomology is similar. $\qquad\qquad\qquad\qquad\qquad \square$

Lemma (Exact sequence for proof of (2.6.1)) *Let A be a ring, B an A-module, C an A-algebra and M a C-module. We have an exact sequence*

$$\mathrm{Tor}_2^A(B, M) \to \mathrm{Tor}_2^C(B \otimes_A C, M) \to$$

$$\mathrm{Tor}_1^A(B, C) \otimes_C M \to \mathrm{Tor}_1^A(B, M) \to \mathrm{Tor}_1^C(B \otimes_A C, M) \to 0.$$

Proof Let $0 \to K \xrightarrow{\lambda} P \xrightarrow{\mu} B \to 0$ be an exact sequence of A-modules with P projective. We have an exact sequence

$$0 \to \operatorname{Tor}_1^A(B,C) \to K \otimes_A C \xrightarrow{\lambda_*} P \otimes_A C \xrightarrow{\mu_*} B \otimes_A C \to 0.$$

Let $D = \ker(\mu_*) = \operatorname{im}(\lambda_*)$. Applying $- \otimes_C M$ to the exact sequences

$$0 \to \operatorname{Tor}_1^A(B,C) \to K \otimes_A C \to D \to 0$$
$$0 \to D \to P \otimes_A C \to B \otimes_A C \to 0$$

and $- \otimes_A M$ to the exact sequence

$$0 \to K \to P \to B \to 0$$

we obtain a commutative diagram with exact rows and columns

$$
\begin{array}{ccc}
 & 0 & 0 \\
 & \downarrow & \downarrow \\
 & \operatorname{Tor}_1^A(B,M) \to \operatorname{Tor}_1^C(B \otimes_A C, M) \\
\operatorname{Tor}_1^C(K \otimes_A C, M) \to & \downarrow & \downarrow \\
\operatorname{Tor}_1^C(D,M) \to \operatorname{Tor}_1^A(B,C) \otimes_C M \to & K \otimes_A M & \to & D \otimes_C M \to 0 \\
 & \downarrow & \downarrow \\
 & P \otimes_A M & = & (P \otimes_A C) \otimes_C M \\
 & \downarrow & \downarrow \\
 & B \otimes_A M & = & (B \otimes_A C) \otimes_C M
\end{array}
$$

Moreover, since $0 \to D \to P \otimes_A C \to B \otimes_A C \to 0$ is a projective presentation over C, we have $\operatorname{Tor}_1^C(D,M) = \operatorname{Tor}_2^C(B \otimes_A C, M)$. So from the above diagram we obtain an exact sequence

$$\operatorname{Tor}_1^C(K \otimes_A C, M) \to \operatorname{Tor}_2^C(B \otimes_A C, M) \to$$
$$\operatorname{Tor}_1^A(B,C) \otimes_C M \to \operatorname{Tor}_1^A(B,M) \to \operatorname{Tor}_1^C(B \otimes_A C, M) \to 0.$$

So it is enough to show an epimorphism

$$\operatorname{Tor}_2^A(B,M) \to \operatorname{Tor}_1^C(K \otimes_A C, M).$$

Since $\operatorname{Tor}_2^A(B,M) = \operatorname{Tor}_1^A(K,M)$, the same reasoning for the A-module K instead of B gives an epimorphism

$$\operatorname{Tor}_1^A(K,M) \to \operatorname{Tor}_1^C(K \otimes_A C, M).$$

$\square$

A similar proof in cohomology gives an exact sequence

$$0 \to \mathrm{Ext}^1_{A/I}(J/IJ, M) \to \mathrm{Ext}^1_A(J, M) \to \mathrm{Hom}_{A/I}(\mathrm{Tor}^A_1(J, A/I), M)$$

for ideals $J \subset I$ of A and an A/I-module M, which in the case $J = I^n$ was used in the proof of (2.3.4).

Bibliography

[An1] M. André. Homologie des algèbres commutatives. Springer, 1974.

[An2] M. André. Localisation de la lissité formelle. Manuscripta Math. 13 (1974), 297-307.

[An3] M. André. Modules des différentielles en caractéristique p. Manuscripta math. 62 (1988), 477-502.

[An4] M. André. Cinq exposés sur la désingularisation. Ecole Polytechnique Fédérale de Lausanne, 1991.

[An5] M. André. Homomorphismes réguliers en caractéristique p. C. R. Acad. Sci. Paris Sér. I Math. 316 (1993), no. 7, 643-646.

[An6] M. André. Autre démonstration du théorème liant régularité et platitude en caractéristique p. Manuscripta Math. 82 (1994), no. 3-4, 363-379.

[AM] M.F. Atiyah, I.G. Macdonald. Introduction to commutative algebra. Addison-Wesley, 1969.

[Av1] L.L. Avramov. Flat morphisms of complete intersections. Doklady Akad. Nauk SSSR 225 (1975), 11-14 (english trans. Soviet Math. Dokl 16 (1975), 1413-1417).

[Av2] L.L. Avramov. Homology of local flat extensions and complete intersection defects. Math. Ann. 228 (1977), 27-37.

[Av3] L.L. Avramov. Descente des déviations par homomorphismes locaux et génération des idéaux de dimension projective finie. C. R. Acad. Sci. Paris Sér. I Math. 295 (1982), no. 12, 665-668.

[BM] A. Blanco, J. Majadas. Sur les morphismes d'intersection complète en caractéristique p. J. of Algebra 208 (1998), 35-42.

[Bo] N. Bourbaki. Algèbre Commutative. Hermann / Masson, Paris.

[Br] A. Brezuleanu. Smoothness and regularity. Compositio Math. 24 (1972), 1-10.

[BDR] A. Brezuleanu, T. Dumitrescu, N. Radu. Local Algebras. Editura Universitatii Bucuresti, 1993.

[BR1] A. Brezuleanu, N. Radu. Sur la localisation de la lissité formelle. C. R. Acad. Sci. Paris 276 (1973), A439-A441.

[BR2] A. Brezuleanu, N. Radu. On the localization of formal smoothness. Rev. Roum. Math. Pures et Appl. 20 (1975), 189-200.

[BR3] A. Brezuleanu, N. Radu. Excellent rings and good separation of the module of differentials. Rev. Roum. Math. Pures et Appl. 23 (1978), 1455-1470.

[EGA] A. Grothendieck. Éléments de Géométrie Algébrique, chap.IV, Première

Partie. Publ. Math. IHES 20, 1964.

[GL] T.H. Gulliksen, G. Levin. Homology of local rings. Queen's Paper in Pure and Applied Mathematics, No. 20 Queen's University, Kingston, Ont. 1969

[Ku] E. Kunz. Characterizations of regular local rings for characteristic p. Amer. J. Math. 91 (1969) 772-784.

[LS] S. Lichtenbaum, M. Schlessinger. The cotangent complex of a morphism. Trans. Amer. Math. Soc. 128 (1967), 41-70.

[Mt] H. Matsumura. Commutative Ring Theory. Cambridge University Press, 1986.

[Og1] T. Ogoma, Noetherian property of inductive limits of Noetherian local rings. Proc. Japan Acad. Ser. A Math. Sci. 67 (1991), no. 3, 68-69.

[Og2] T. Ogoma. General Néron desingularization based on the idea of Popescu. J. Algebra 167 (1994), 57-84.

[Po1] D. Popescu. General Néron desingularization. Nagoya Math. J. 100 (1985), 97-126.

[Po2] D. Popescu. General Néron desingularization and approximation. Nagoya Math. J. 104 (1986), 85-115.

[Po3] D. Popescu. Letter to the editor: "General Néron desingularization and approximation" [Nagoya Math. J. 104 (1986), 85-115]. Nagoya Math. J. 118 (1990), 45-53.

[Qu] D. Quillen. On the (co-)homology of commutative rings. Proc. Sympos. Pure Math. 17 (1970) 65-87.

[Ra1] N. Radu. Sur un critère de lissité formelle. Bull. Math. Soc, Sci Math. Roumaine 21 (1977), 133-135.

[Ra2] N. Radu. Sur la structure des algèbres locales noethériennes formellement lisses. Analele Univ. Bucuresti Mathematica 29 (1980), 81-84.

[Ra3] N. Radu. Une classe d'anneaux noethériens. Rev. Roumaine Math. Pures Appl. 37 (1992), no. 1, 79-82.

[Sp] M. Spivakovsky. A new proof of D. Popescu's theorem on smoothing of ring homomorphisms. J. Amer. Math. Soc. 12 (1999), 381–444.

[Sw] R.G. Swan. Néron-Popescu desingularization. Algebra and geometry (Taipei, 1995), 135-192, Lect. Algebra Geom., 2, Internat. Press, Cambridge, MA, 1998.

[ZS] O. Zariski, P. Samuel. Commutative Algebra, Springer, 1975.

Other works related to parts of this book:

[An7] M. André. Algèbres graduées associées et algèbres symétriques plates. Comment. Math. Helv. 49 (1974), 277-301.

[An8] M. André. La $(2p+1)$-ème déviation d'un anneau local. Enseignement Math. (2) 23 (1977), no. 3-4, 239-248.

[An9] M. André. Produits de Massey et $(2p+1)$-èmes déviations. Séminaire d'Algèbre Paul Dubreil et Marie-Paule Malliavin, 32ème année (Paris, 1979), pp. 341-359, Lecture Notes in Math., 795, Springer, Berlin, 1980.

[An10] M. André. Le caractère additif des déviations des anneaux locaux. Comment. Math. Helv. 57 (1982), no. 4, 648-675.

[Ar] M. Artin. Algebraic approximation of structures over complete local rings. Inst. Hautes Études Sci. Publ. Math. No. 36 (1969), 23-58.

[As] E.F. Assmus, Jr. On the homology of local rings. Illinois J. Math. 3 (1959), 187-199.

[Av4] L.L. Avramov. Local rings of finite simplicial dimension. Bull. Amer. Math. Soc. (N.S.) 10 (1984), no. 2, 289-291.

[Av5] L.L. Avramov. Locally complete intersection homomorphisms and a conjecture of Quillen on the vanishing of cotangent homology. Ann. of Math. (2) 150 (1999), no. 2, 455-487.

[AFH] L.L. Avramov, H.-B. Foxby, B. Herzog. Structure of local homomorphisms. J. Algebra 164 (1994), no. 1, 124-145.

[AHa] L.L. Avramov, S. Halperin. On the nonvanishing of cotangent cohomology. Comment. Math. Helv. 62 (1987), no. 2, 169-184.

[AHe] L.L. Avramov, J. Herzog. Jacobian criteria for complete intersections. The graded case. Invent. Math. 117 (1994), no. 1, 75-88.

[AI] L.L. Avramov, S. Iyengar. André–Quillen homology of algebra retracts. Ann. Sci. École Norm. Sup. (4) 36 (2003), no. 3, 431-462.

[AMi] L.L. Avramov, C. Miller. Frobenius powers of complete intersections. Math. Res. Lett. 8 (2001), no. 1-2, 225-232.

[BMR] A. Blanco, J. Majadas, A.G. Rodicio. On the acyclicity of the Tate complex. J. Pure Appl. Algebra 131 (1998), no. 2, 125-132.

[BR4] A. Brezuleanu, N. Radu. On the Jacobian criterion of Nagata-Grothendieck. Rev. Roumaine Math. Pures Appl. 31 (1986), no. 6, 513-517.

[BR5] A. Brezuleanu, N. Radu. Stability of geometric regularity as consequence of Nagata's Jacobian criterion. Stud. Cerc. Mat. 40 (1988), no. 6, 457-462.

[BH] W. Bruns, J. Herzog. Cohen-Macaulay rings. Cambridge Studies in Advanced Mathematics, 39. Cambridge University Press, Cambridge, 1993.

[Co] I.S. Cohen. On the structure and ideal theory of complete local rings. Trans. Amer. Math. Soc. 59, (1946). 54-106.

[Du1] T. Dumitrescu. On a theorem of N. Radu and André. Stud. Cerc. Mat. 46 (1994), no. 4, 445-447.

[Du2] T. Dumitrescu. Reducedness, formal smoothness and approximation in characteristic p. Comm. Algebra 23 (1995), no. 5, 1787-1795.

[Du3] T. Dumitrescu. Regularity and finite flat dimension in characteristic $p > 0$. Comm. Algebra 24 (1996), no. 10, 3387-3401.

[DR] T. Dumitrescu, N. Radu. On a theorem of A. Grothendieck. Stud. Cerc. Mat. 48 (1996), no. 5-6, 319-323.

[Fe] D. Ferrand. Suite régulière et intersection complète. C. R. Acad. Sci. Paris Sér. A-B 264 1967 A427-A428.

[FR] L. Franco, A.G. Rodicio. On the vanishing of the second André–Quillen homology of a local homomorphism. J. Algebra 155 (1993), no. 1, 137-141.

[G] A. Geddes. On the embedding theorems for complete local rings. Proc. London Math. Soc. (3) 6 (1956), 343-354.

[He] J. Herzog. Homological properties of the module of differentials. Atas VI Escola de Algebra, IMPA, Rio de Janeiro, 1980, 33-64.

[Ka] I. Kaplansky. Commutative rings. Revised edition. The University of Chicago Press, Chicago, Ill.-London, 1974.

[Ku1] E. Kunz. Kähler differentials. Advanced Lectures in Mathematics. Friedr. Vieweg & Sohn, Braunschweig, 1986.

[La] D. Lazard. Autour de la platitude. Bull. Soc. Math. France 97 (1969) 81-128.

[Li] J. Lipman. Free derivation modules on algebraic varieties. Amer. J. Math. 87 (1965) 874-898.

[Mr]   J. Marot. Sur les homomorphismes d'intersection complète. C. R. Acad. Sci. Paris Sér. I Math. 294 (1982), no. 12, 381-384.

[Mt1]   H. Matsumura. Commutative algebra. Second edition. Mathematics Lecture Note Series, 56. Benjamin/Cummings Publishing Co., Inc., Reading, Mass., 1980.

[N1]   M. Nagata. On the structure of complete local rings. Nagoya Math. J. 1 (1950), 63-70; 5 (1953), 145-147.

[N2]   M. Nagata. Local rings. Interscience Tracts in Pure and Applied Mathematics, No. 13. Interscience Publishers, a division of John Wiley & Sons, New York-London, 1962.

[Na]   M. Narita. On the structure of complete local rings. J. Math. Soc. Japan 7 (1955), 435-443.

[NP]   V. Nica, D. Popescu. A structure theorem on formally smooth morphisms in positive characteristic. J. Algebra 100 (1986), no. 2, 436-455.

[PS]   C. Peskine, L. Szpiro. Dimension projective finie et cohomologie locale. Applications à la démonstration de conjectures de M. Auslander, H. Bass et A. Grothendieck. Inst. Hautes Études Sci. Publ. Math. No. 42 (1973), 47-119.

[Qu1]   D. Quillen. Homology of Commutative Rings. MIT mimeographed notes, 1967.

[Ra4]   N. Radu. Sur les algèbres dont le module des différentielles est plat. Rev. Roumaine Math. Pures Appl. 21 (1976), no. 7, 933-939.

[Rg]   A. Ragusa. On openness of H_n-locus and semicontinuity of nth deviation. Proc. Amer. Math. Soc. 80 (1980), no. 2, 201-209.

[RG]   M. Raynaud, L. Gruson. Critères de platitude et de projectivité. Techniques de "platification" d'un module. Invent. Math. 13 (1971), 1-89.

[Ro1]   A.G. Rodicio. On the free character of the first koszul homology module. J. Pure Appl. Algebra 80 (1992), 59–64.

[Ro2]   A.G. Rodicio. Flat exterior Tor algebras and cotangent complexes. Comment. Math. Helv. 70 (1995), 546–557. Erratum: Comment. Math. Helv. 71 (1996), 338.

[Se]   J.-P. Serre. Algèbre locale. Multiplicités. Cours au Collège de France, 1957-1958, rédigé par Pierre Gabriel. Seconde édition, 1965. Lecture Notes in Mathematics, 11. Springer-Verlag, Berlin-New York, 1965.

[Ta]   M. Tabaâ. Sur les homomorphismes d'intersection complète. C. R. Acad. Sci. Paris Sér. I Math. 298 (1984), no. 18, 437-439.

[Tn1]   H. Tanimoto. Some characterizations of smoothness. J. Math. Kyoto Univ. 23 (1983), no. 4, 695-706.

[Tn2]   H. Tanimoto. Smoothness of Noetherian rings. Nagoya Math. J. 95 (1984), 163-179.

[Tt]   J. Tate. Homology of Noetherian rings and local rings. Illinois J. Math. 1 (1957), 14-27.

[Ul]   B. Ulrich. Vanishing of cotangent functors. Math. Z. 196 (1987), no. 4, 463-484.

[Va]   W. Vasconcelos. V. Ideals generated by R-sequences. J. Algebra 6 (1967) 309-316.

Index

Printed in the United States
by Baker & Taylor Publisher Services